PROBLEMS
AND
SOLUTIONS
IN
INTRODUCTORY
AND
ADVANCED
MATRIX CALCULUS

T0331570

Willi-Hans Steeb

University of Johannesburg, South Africa

PROBLEMS
AND
SOLUTIONS
IN
INTRODUCTORY
AND
ADVANCED
MATRIX CALCULUS

World Scientific

NEW JERSEY · LONDON · SINGAPORE · BEIJING · SHANGHAI · HONG KONG · TAIPEI · CHENNAI

Published by

World Scientific Publishing Co. Pte. Ltd.
5 Toh Tuck Link, Singapore 596224
USA office: 27 Warren Street, Suite 401-402, Hackensack, NJ 07601
UK office: 57 Shelton Street, Covent Garden, London WC2H 9HE

Library of Congress Cataloging-in-Publication Data
Steeb, W.-H.
    Problems and solutions in introductory and advanced matrix calculus / Willi-Hans Steeb.
      p. cm.
    Includes bibliographical references and index.
    ISBN-13 978-981-256-916-5 (alk. paper) -- ISBN-13 978-981-270-202-9 (pbk: alk. paper)
    ISBN-10 981-256-916-2 (alk. paper) -- ISBN-10 981-270-202-4 (pbk: alk. paper)
    1. Matrices--Problems, exercises, etc. 2. Calculus. 3. Mathematical physics. I. Steeb, W.-H.
II. Title.

    QA188.S664 2006
    512.9'434--dc22   2006047621

British Library Cataloguing-in-Publication Data
A catalogue record for this book is available from the British Library.

Printed in Singapore

Preface

The purpose of this book is to supply a collection of problems in introductory and advanced matrix problems together with their detailed solutions which will prove to be valuable to undergraduate and graduate students as well as to research workers in these fields. Each chapter contains an introduction with the essential definitions and explanations to tackle the problems in the chapter. If necessary, other concepts are explained directly with the present problems. Thus the material in the book is self-contained. The topics range in difficulty from elementary to advanced. Students can learn important principles and strategies required for problem solving. Lecturers will also find this text useful either as a supplement or text, since important concepts and techniques are developed in the problems.

A large number of problems are related to applications. Applications include wavelets, linear integral equations, Kirchhoff's laws, global positioning systems, Floquet theory, octonians, random walks, Kronecker product and images. A number of problems useful in quantum physics and graph theory are also provided. Advanced topics include groups and matrices, Lie groups and matrices and Lie algebras and matrices. Exercises for matrix-valued differential forms are also included.

The book can also be used as a text for linear and multilinear algebra or matrix theory. The material was tested in my lectures given around the world.

The International School for Scientific Computing (ISSC) provides certificate courses for this subject. Please contact the author if you want to do this course or other courses of the ISSC.

e-mail addresses of the author:

steebwilli@gmail.com
steeb_wh@yahoo.com

Home page of the author:

http://issc.uj.ac.za

Contents

Notation

:=	is defined as
$\in$	belongs to (a set)
$\notin$	does not belong to (a set)
$\cap$	intersection of sets
$\cup$	union of sets
$\emptyset$	empty set
N	set of natural numbers
Z	set of integers
Q	set of rational numbers
R	set of real numbers
$\mathbf{R}^+$	set of nonnegative real numbers
C	set of complex numbers
$\mathbf{R}^n$	n-dimensional Euclidean space
	space of column vectors with n real components
$\mathbf{C}^n$	n-dimensional complex linear space
	space of column vectors with n complex components
$\mathcal{H}$	Hilbert space
i	$\sqrt{-1}$
$\Re z$	real part of the complex number z
$\Im z$	imaginary part of the complex number z
$|z|$	modulus of complex number z
	$|x + iy| = (x^2 + y^2)^{1/2}$, $x, y \in \mathbf{R}$
$T \subset S$	subset T of set S
$S \cap T$	the intersection of the sets S and T
$S \cup T$	the union of the sets S and T
$f(S)$	image of set S under mapping f
$f \circ g$	composition of two mappings $(f \circ g)(x) = f(g(x))$
$\mathbf{x}$	column vector in $\mathbf{C}^n$
$\mathbf{x}^T$	transpose of $\mathbf{x}$ (row vector)
$\mathbf{0}$	zero (column) vector

$\|\cdot\|$	norm
$\mathbf{x} \cdot \mathbf{y} \equiv \mathbf{x}^*\mathbf{y}$	scalar product (inner product) in $\mathbf{C}^n$
$\mathbf{x} \times \mathbf{y}$	vector product in $\mathbf{R}^3$
A, B, C	$m \times n$ matrices
$\det(A)$	determinant of a square matrix A
$\text{tr}(A)$	trace of a square matrix A
$\text{rank}(A)$	rank of matrix A
A^T	transpose of matrix A
$\overline{A}$	conjugate of matrix A
A^*	conjugate transpose of matrix A
$A^\dagger$	conjugate transpose of matrix A
	(notation used in physics)
A^{-1}	inverse of square matrix A (if it exists)
I_n	$n \times n$ unit matrix
I	unit operator
0_n	$n \times n$ zero matrix
AB	matrix product of $m \times n$ matrix A
	and $n \times p$ matrix B
$A \bullet B$	Hadamard product (entry-wise product)
	of $m \times n$ matrices A and B
$[A, B] := AB - BA$	commutator for square matrices A and B
$[A, B]_+ := AB + BA$	anticommutator for square matrices A and B
$A \otimes B$	Kronecker product of matrices A and B
$A \oplus B$	Direct sum of matrices A and B
δ_{jk}	Kronecker delta with $\delta_{jk} = 1$ for $j = k$
	and $\delta_{jk} = 0$ for $j \neq k$
λ	eigenvalue
ϵ	real parameter
t	time variable
$\hat{H}$	Hamilton operator

The Pauli spin matrices are used extensively in the book. They are given
by

$$\sigma_x := \begin{pmatrix} 0 & 1 \\ 1 & 0 \end{pmatrix}, \quad \sigma_y := \begin{pmatrix} 0 & -i \\ i & 0 \end{pmatrix}, \quad \sigma_z := \begin{pmatrix} 1 & 0 \\ 0 & -1 \end{pmatrix}.$$

In some cases we will also use σ_1, σ_2 and σ_3 to denote σ_x, σ_y and σ_z .

Chapter 1

Basic Operations

Let $\mathcal{F}$ be a field, for example the set of real numbers $\mathbf{R}$ or the set of complex numbers $\mathbf{C}$. Let m, n be two integers ≥ 1. An array A of numbers in $\mathcal{F}$

$$\begin{pmatrix} a_{11} & a_{12} & a_{13} & \cdots & a_{1n} \\ a_{21} & a_{22} & a_{23} & \cdots & a_{2n} \\ \vdots & \vdots & \vdots & \ddots & \vdots \\ a_{m1} & a_{m2} & a_{m3} & \cdots & a_{mn} \end{pmatrix} = (a_{ij})$$

is called an $m \times n$ *matrix* with entry a_{ij} in the ith row and jth column. A *row vector* is a $1 \times n$ matrix. A *column vector* is an $n \times 1$ matrix. We have a *zero matrix*, in which $a_{ij} = 0$ for all i, j.

Let $A = (a_{ij})$ and $B = (b_{ij})$ be two $m \times n$ matrices. We define $A + B$ to be the $m \times n$ matrix whose entry in the i-th row and j-th column is $a_{ij} + b_{ij}$. Matrix multiplication is only defined between two matrices if the number of columns of the first matrix is the same as the number of rows of the second matrix. If A is an $m \times n$ matrix and B is an $n \times p$ matrix, then the matrix product AB is an $m \times p$ matrix defined by

$$(AB)_{ij} = \sum_{r=1}^{n} a_{ir} b_{rj}$$

for each pair i and j, where $(AB)_{ij}$ denotes the (i, j)th entry in AB. Let $A = (a_{ij})$ and $B = (b_{ij})$ be two $m \times n$ matrices with entries in some field. Then their Hadamard product is the entry-wise product of A and B, that is the $m \times n$ matrix $A \bullet B$ whose (i, j)th entry is $a_{ij} b_{ij}$.

Problem 1.   Let $\mathbf{x}$ be a column vector in $\mathbf{R}^n$ and $\mathbf{x} \neq \mathbf{0}$. Let

$$A = \frac{\mathbf{x}\mathbf{x}^T}{\mathbf{x}^T\mathbf{x}}$$

where T denotes the transpose, i.e. $\mathbf{x}^T$ is a row vector. Calculate A^2.

Solution 1.   Obviously $\mathbf{x}\mathbf{x}^T$ is a nonzero $n \times n$ matrix and $\mathbf{x}^T\mathbf{x}$ is a nonzero real number. We find

$$
\begin{aligned}
A^2 &= \left(\frac{\mathbf{x}\mathbf{x}^T}{\mathbf{x}^T\mathbf{x}}\right)^2 \\
&= \frac{(\mathbf{x}\mathbf{x}^T)(\mathbf{x}\mathbf{x}^T)}{(\mathbf{x}^T\mathbf{x})^2} \\
&= \frac{\mathbf{x}(\mathbf{x}^T\mathbf{x})\mathbf{x}^T}{(\mathbf{x}^T\mathbf{x})^2} \\
&= \frac{\mathbf{x}\mathbf{x}^T}{\mathbf{x}^T\mathbf{x}} \\
&= A
\end{aligned}
$$

where we used the fact that matrix multiplication is *associative*. Thus $A^2 = A$.

Problem 2.   Consider the vector in $\mathbf{R}^8$

$$\mathbf{x}^T = (20.0\ 6.0\ 4.0\ 2.0\ 10.0\ 6.0\ 8.0\ 4.0)$$

where T denotes transpose. Consider the matrices

$$H_1 = \frac{1}{2}\begin{pmatrix} 1 & 1 & 0 & 0 & 0 & 0 & 0 & 0 \\ 0 & 0 & 1 & 1 & 0 & 0 & 0 & 0 \\ 0 & 0 & 0 & 0 & 1 & 1 & 0 & 0 \\ 0 & 0 & 0 & 0 & 0 & 0 & 1 & 1 \end{pmatrix}$$

$$G_1 = \frac{1}{2}\begin{pmatrix} 1 & -1 & 0 & 0 & 0 & 0 & 0 & 0 \\ 0 & 0 & 1 & -1 & 0 & 0 & 0 & 0 \\ 0 & 0 & 0 & 0 & 1 & -1 & 0 & 0 \\ 0 & 0 & 0 & 0 & 0 & 0 & 1 & -1 \end{pmatrix}$$

$$H_2 = \frac{1}{2}\begin{pmatrix} 1 & 1 & 0 & 0 \\ 0 & 0 & 1 & 1 \end{pmatrix}, \qquad G_2 = \frac{1}{2}\begin{pmatrix} 1 & -1 & 0 & 0 \\ 0 & 0 & 1 & -1 \end{pmatrix}$$

$$H_3 = \frac{1}{2}(1\ \ 1), \qquad G_3 = \frac{1}{2}(1\ \ -1).$$

(i) Calculate
$$H_1\mathbf{x}, \qquad G_1\mathbf{x}$$
$$H_2H_1\mathbf{x}, \quad G_2H_1\mathbf{x}, \quad H_2G_1\mathbf{x}, \quad G_2G_1\mathbf{x}$$
$$H_3H_2H_1\mathbf{x}, \quad G_3H_2H_1\mathbf{x}, \quad H_3G_2H_1\mathbf{x}, \quad G_3G_2H_1\mathbf{x},$$
$$H_3H_2G_1\mathbf{x}, \quad G_3H_2G_1\mathbf{x}, \quad H_3G_2G_1\mathbf{x}, \quad G_3G_2G_1\mathbf{x}$$

(ii) Calculate
$$H_jH_j^T, \quad G_jG_j^T, \quad H_jG_j^T$$

for $j = 1, 2, 3$.

(iii) How can we reconstruct the original vector $\mathbf{x}$ from the vector

$$(H_3H_2H_1\mathbf{x}, G_3H_2H_1\mathbf{x}, H_3G_2H_1\mathbf{x}, G_3G_2H_1\mathbf{x}, H_3H_2G_1\mathbf{x}, G_3H_2G_1\mathbf{x},$$
$$H_3G_2G_1\mathbf{x}, G_3G_2G_1\mathbf{x})$$

The problem plays a role in *wavelet theory*.

Solution 2.   (i) We find

$$H_1\mathbf{x} = \begin{pmatrix} 13.0 \\ 3.0 \\ 8.0 \\ 6.0 \end{pmatrix}, \qquad G_1\mathbf{x} = \begin{pmatrix} 7.0 \\ 1.0 \\ 2.0 \\ 2.0 \end{pmatrix}.$$

Thus we have the vector

$$(13.0 \ 3.0 \ 8.0 \ 6.0 \ 7.0 \ 1.0 \ 2.0 \ 2.0)^T.$$

Next we find

$$H_2H_1\mathbf{x} = \begin{pmatrix} 8.0 \\ 7.0 \end{pmatrix}, \quad G_2H_1\mathbf{x} = \begin{pmatrix} 5.0 \\ 1.0 \end{pmatrix},$$

$$H_2G_1\mathbf{x} = \begin{pmatrix} 4.0 \\ 2.0 \end{pmatrix}, \quad G_2G_1\mathbf{x} = \begin{pmatrix} 3.0 \\ 0.0 \end{pmatrix}.$$

Thus we have the vector

$$(8.0 \ 7.0 \ 5.0 \ 1.0 \ 4.0 \ 2.0 \ 3.0 \ 0.0)^T.$$

Finally we have

$$H_3H_2H_1\mathbf{x} = 7.5, \quad G_3H_2H_1\mathbf{x} = 0.5, \quad H_3G_2H_1\mathbf{x} = 3.0, \quad G_3G_2H_1\mathbf{x} = 2.0$$

$$H_3H_2G_1\mathbf{x} = 3.0, \quad G_3H_2G_1\mathbf{x} = 1.0, \quad H_3G_2G_1\mathbf{x} = 1.5, \quad G_3G_2G_1\mathbf{x} = 1.5.$$

Thus we obtain the vector

$$(7.5 \ 0.5 \ 3.0 \ 2.0 \ 3.0 \ 1.0 \ 1.5 \ 1.5).$$

(ii) We find

$$H_1 H_1^T = \frac{1}{2}I_4, \quad G_1 G_1^T = \frac{1}{2}I_4, \quad H_1 G_1^T = 0_4$$

$$H_2 H_2^T = \frac{1}{2}I_2, \quad G_2 G_2^T = \frac{1}{2}I_2, \quad H_2 G_2^T = 0_2$$

$$H_3 H_3^T = \frac{1}{2}, \quad G_3 G_3^T = \frac{1}{2}, \quad H_3 G_3^T = 0$$

where 0_4 is the 4×4 zero matrix and 0_2 is the 2×2 zero matrix.

(iii) Let

$$\mathbf{w} = (7.5 \ 0.5 \ 3.0 \ 2.0 \ 3.0 \ 1.0 \ 1.5 \ 1.5)^T$$

Consider the matrices

$$X_1 = \begin{pmatrix} 1 & 1 & 0 & 0 & 0 & 0 & 0 & 0 \\ 1 & -1 & 0 & 0 & 0 & 0 & 0 & 0 \\ 0 & 0 & 1 & 1 & 0 & 0 & 0 & 0 \\ 0 & 0 & 1 & -1 & 0 & 0 & 0 & 0 \end{pmatrix}$$

$$Y_1 = \begin{pmatrix} 0 & 0 & 0 & 0 & 1 & 1 & 0 & 0 \\ 0 & 0 & 0 & 0 & 1 & -1 & 0 & 0 \\ 0 & 0 & 0 & 0 & 0 & 0 & 1 & 1 \\ 0 & 0 & 0 & 0 & 0 & 0 & 1 & -1 \end{pmatrix}.$$

Then

$$X_1 \mathbf{w} = \begin{pmatrix} 8 \\ 7 \\ 5 \\ 1 \end{pmatrix}, \quad Y_1 \mathbf{w} = \begin{pmatrix} 4 \\ 2 \\ 3 \\ 0 \end{pmatrix}.$$

Now let

$$X_2 = Y_2 = \begin{pmatrix} 1 & 0 & 1 & 0 \\ 1 & 0 & -1 & 0 \\ 0 & 1 & 0 & 1 \\ 0 & 1 & 0 & -1 \end{pmatrix}.$$

Then

$$X_2(X_1 \mathbf{w}) = \begin{pmatrix} 13 \\ 3 \\ 8 \\ 6 \end{pmatrix}, \quad Y_2(Y_1 \mathbf{w}) = \begin{pmatrix} 7 \\ 1 \\ 2 \\ 2 \end{pmatrix}.$$

Thus the original vector $\mathbf{x}$ is reconstructed from

$$\begin{pmatrix} X_2(X_1\mathbf{w}) \\ Y_2(Y_1\mathbf{w}) \end{pmatrix}.$$

The odd entries come from $X_2(X_1\mathbf{w}) + Y_2(Y_1\mathbf{w})$ and the even ones from $X_2(X_1\mathbf{w}) - Y_2(Y_1\mathbf{w})$.

Problem 3. Consider the 8×8 *Hadamard matrix*

$$H = \begin{pmatrix} 1 & 1 & 1 & 1 & 1 & 1 & 1 & 1 \\ 1 & 1 & 1 & 1 & -1 & -1 & -1 & -1 \\ 1 & 1 & -1 & -1 & -1 & -1 & 1 & 1 \\ 1 & 1 & -1 & -1 & 1 & 1 & -1 & -1 \\ 1 & -1 & -1 & 1 & 1 & -1 & -1 & 1 \\ 1 & -1 & -1 & 1 & -1 & 1 & 1 & -1 \\ 1 & -1 & 1 & -1 & -1 & 1 & -1 & 1 \\ 1 & -1 & 1 & -1 & 1 & -1 & 1 & -1 \end{pmatrix}.$$

(i) Do the 8 column vectors in the matrix H form a basis in $\mathbf{R}^8$? Prove or disprove.
(ii) Calculate HH^T, where T denotes transpose. Compare the results from (i) and (ii) and discuss.

Solution 3. (i) Calculating the scalar product of the column vectors (and obviously also the row vectors) we find that they are pairwise orthogonal to each other. Since they all nonzero vectors we have a basis in $\mathbf{R}^8$, which is not normalized.
(ii) We find

$$HH^T = 8I_8.$$

Thus the matrix H is invertible and the column or row vectors must form a basis. Since the right-hand side is $8I_8$ the basis must be orthogonal.

Problem 4. Show that any 2×2 complex matrix has a unique representation of the form

$$a_0 I_2 + ia_1\sigma_1 + ia_2\sigma_2 + ia_3\sigma_3$$

for some $a_0, a_1, a_2, a_3 \in \mathbf{C}$, where I_2 is the 2×2 identity matrix and $\sigma_1, \sigma_2, \sigma_3$ are the *Pauli spin matrices*

$$\sigma_1 := \begin{pmatrix} 0 & 1 \\ 1 & 0 \end{pmatrix}, \qquad \sigma_2 := \begin{pmatrix} 0 & -i \\ i & 0 \end{pmatrix}, \qquad \sigma_3 := \begin{pmatrix} 1 & 0 \\ 0 & -1 \end{pmatrix}.$$

Solution 4. Since

$$a_0 I_2 + ia_1\sigma_1 + ia_2\sigma_2 + ia_3\sigma_3 = \begin{pmatrix} a_0 + ia_3 & ia_1 + a_2 \\ ia_1 - a_2 & a_0 - ia_3 \end{pmatrix}$$

we obtain

$$a_0 I_2 + ia_1\sigma_1 + ia_2\sigma_2 + ia_3\sigma_3 = \begin{pmatrix} \alpha & \beta \\ \gamma & \delta \end{pmatrix}$$

where $\alpha, \beta, \gamma, \delta \in \mathbf{C}$. Thus

$$a_0 = \frac{\alpha + \delta}{2}, \qquad a_1 = \frac{\beta + \gamma}{2i}, \qquad a_2 = \frac{\beta - \gamma}{2}, \qquad a_3 = \frac{\alpha - \delta}{2i}.$$

Problem 5. Let A, B be $n \times n$ matrices such that $ABAB = 0_n$. Can we conclude that $BABA = 0_n$?

Solution 5. There are no 1×1 or 2×2 counterexamples. For 2×2 matrices $ABAB = 0_n$ implies $B(ABAB)A = 0_n$. Therefore the matrix BA is *nilpotent*, i.e. $(BA)^3 = 0_n$. If a 2×2 matrix C is nilpotent, its characteristic polynomial is λ^2 and therefore $C^2 = 0_n$ by the *Cayley-Hamilton theorem*. Thus $BABA = 0_n$. For $n = 3$ we find the counterexample

$$A = \begin{pmatrix} 0 & 0 & 1 \\ 0 & 0 & 0 \\ 0 & 1 & 0 \end{pmatrix}, \qquad B = \begin{pmatrix} 0 & 0 & 1 \\ 1 & 0 & 0 \\ 0 & 0 & 0 \end{pmatrix}$$

with

$$BABA = \begin{pmatrix} 0 & 0 & 1 \\ 0 & 0 & 0 \\ 0 & 0 & 0 \end{pmatrix}.$$

Problem 6. A square matrix A over $\mathbf{C}$ is called *skew-hermitian* if $A = -A^*$. Show that such a matrix is *normal*, i.e., we have $AA^* = A^*A$.

Solution 6. We have

$$AA^* = -A^*A^* = (-A^*)(-A) = A^*A.$$

Problem 7. Let A be an $n \times n$ skew-hermitian matrix over $\mathbf{C}$, i.e. $A^* = -A$. Let U be an $n \times n$ *unitary matrix*, i.e., $U^* = U^{-1}$. Show that $B := U^*AU$ is a skew-hermitian matrix.

Solution 7. We have $A^* = -A$. Thus from $B = U^*AU$ and $U^{**} = U$ it follows that

$$B^* = (U^*AU)^* = U^*A^*U = U^*(-A)U = -U^*AU = -B.$$

Problem 8. Let A, X, Y be $n \times n$ matrices. Assume that

$$XA = I_n, \qquad AY = I_n$$

where I_n is the $n \times n$ unit matrix. Show that $X = Y$.

Solution 8. We have

$$X = XI_n = X(AY) = (XA)Y = I_nY = Y.$$

Problem 9. Let A, B be $n \times n$ matrices. Assume that A is nonsingular, i.e. A^{-1} exists. Show that if $BA = 0_n$, then $B = 0_n$.

Solution 9. We have

$$B = BI_n = B(AA^{-1}) = (BA)A^{-1} = 0_n A^{-1} = 0_n .$$

Problem 10. Let A, B be $n \times n$ matrices and

$$A + B = I_n, \qquad AB = 0_n .$$

Show that $A^2 = A$ and $B^2 = B$.

Solution 10. Multiplying $A + B = I_n$ with A we obtain $A^2 + AB = A$ and therefore $A^2 = A$. Multiplying $A + B = I_n$ with B yields $AB + B^2 = B$ and therefore $B^2 = B$.

Problem 11. Consider the normalized vectors in $\mathbf{R}^2$

$$\begin{pmatrix} \cos \theta_1 \\ \sin \theta_1 \end{pmatrix}, \qquad \begin{pmatrix} \cos \theta_2 \\ \sin \theta_2 \end{pmatrix} .$$

Find the condition on θ_1 and θ_2 such that

$$\begin{pmatrix} \cos \theta_1 \\ \sin \theta_1 \end{pmatrix} + \begin{pmatrix} \cos \theta_2 \\ \sin \theta_2 \end{pmatrix}$$

is normalized. A vector $\mathbf{x} \in \mathbf{R}^n$ is called *normalized* if $\|\mathbf{x}\| = 1$, where $\| \ \|$ denotes the Euclidean norm.

Solution 11. From the condition that the vector

$$\begin{pmatrix} \cos \theta_1 + \cos \theta_2 \\ \sin \theta_1 + \sin \theta_2 \end{pmatrix}$$

is normalized it follows that

$$(\sin \theta_1 + \sin \theta_2)^2 + (\cos \theta_1 + \cos \theta_2)^2 = 1 .$$

Thus we have

$$\sin \theta_1 \sin \theta_2 + \cos \theta_1 \cos \theta_2 = -\frac{1}{2} .$$

It follows that

$$\cos(\theta_1 - \theta_2) = -\frac{1}{2} .$$

Therefore, $\theta_1 - \theta_2 = 2\pi/3$ or $\theta_1 - \theta_2 = 4\pi/3$.

Problem 12.   Let

$$A := \mathbf{x}\mathbf{x}^T + \mathbf{y}\mathbf{y}^T \tag{1}$$

where

$$\mathbf{x} = \begin{pmatrix} \cos\theta \\ \sin\theta \end{pmatrix}, \qquad \mathbf{y} = \begin{pmatrix} \sin\theta \\ -\cos\theta \end{pmatrix}$$

and $\theta \in \mathbf{R}$. Find the matrix A.

Solution 12.   We find

$$A = \begin{pmatrix} \cos^2\theta & \cos\theta\sin\theta \\ \cos\theta\sin\theta & \sin^2\theta \end{pmatrix} + \begin{pmatrix} \sin^2\theta & -\cos\theta\sin\theta \\ -\cos\theta\sin\theta & \cos^2\theta \end{pmatrix} = \begin{pmatrix} 1 & 0 \\ 0 & 1 \end{pmatrix}.$$

The vectors $\mathbf{x}$ and $\mathbf{y}$ form a basis in $\mathbf{R}^2$. Equation (1) is called the *completeness relation*.

Problem 13.   Find a 2×2 matrix A over $\mathbf{R}$ such that

$$A \begin{pmatrix} 1 \\ 0 \end{pmatrix} = \frac{1}{\sqrt{2}} \begin{pmatrix} 1 \\ 1 \end{pmatrix}, \qquad A \begin{pmatrix} 0 \\ 1 \end{pmatrix} = \frac{1}{\sqrt{2}} \begin{pmatrix} 1 \\ -1 \end{pmatrix}.$$

Solution 13.   We find

$$a_{11} = a_{12} = a_{21} = \frac{1}{\sqrt{2}}, \qquad a_{22} = -\frac{1}{\sqrt{2}}.$$

Thus

$$A = \frac{1}{\sqrt{2}} \begin{pmatrix} 1 & 1 \\ 1 & -1 \end{pmatrix}.$$

This matrix is a Hadamard matrix.

Problem 14.   Consider the 2×2 matrix over the complex numbers

$$\Pi(\mathbf{n}) := \frac{1}{2}\left(I_2 + \sum_{j=1}^{3} n_j\sigma_j \right)$$

where $\mathbf{n} := (n_1, n_2, n_3)$ ($n_j \in \mathbf{R}$) is a unit vector, i.e., $n_1^2 + n_2^2 + n_3^2 = 1$. Here σ_1, σ_2, σ_3 are the *Pauli matrices*

$$\sigma_1 = \begin{pmatrix} 0 & 1 \\ 1 & 0 \end{pmatrix}, \qquad \sigma_2 = \begin{pmatrix} 0 & -i \\ i & 0 \end{pmatrix}, \qquad \sigma_3 = \begin{pmatrix} 1 & 0 \\ 0 & -1 \end{pmatrix}$$

and I_2 is the 2×2 unit matrix.

(i) Describe the property of $\Pi(\mathbf{n})$, i.e., find $\Pi^*(\mathbf{n})$, $\mathrm{tr}(\Pi(\mathbf{n}))$ and $\Pi^2(\mathbf{n})$, where tr denotes the trace. The trace is the sum of the diagonal elements of a square matrix.

(ii) Find the vector

$$\Pi(\mathbf{n}) \begin{pmatrix} e^{i\phi}\cos\theta \\ \sin\theta \end{pmatrix} .$$

Discuss.

Solution 14.   (i) For the Pauli matrices we have $\sigma_1^* = \sigma_1$, $\sigma_2^* = \sigma_2$, $\sigma_3^* = \sigma_3$. Thus $\Pi(\mathbf{n}) = \Pi^*(\mathbf{n})$. Since

$$\mathrm{tr}\sigma_1 = \mathrm{tr}\sigma_2 = \mathrm{tr}\sigma_3 = 0$$

and the trace operation is linear, we obtain $\mathrm{tr}(\Pi(\mathbf{n})) = 1$. Since

$$\sigma_1^2 = \sigma_2^2 = \sigma_3^2 = I_2$$

and

$$\sigma_1\sigma_2 + \sigma_2\sigma_1 = 0_2, \quad \sigma_2\sigma_3 + \sigma_3\sigma_2 = 0_2, \quad \sigma_3\sigma_1 + \sigma_1\sigma_3 = 0_2$$

the expression

$$\Pi^2(\mathbf{n}) = \frac{1}{4}\left(I_2 + \sum_{j=1}^{3} n_j\sigma_j\right)^2 = \frac{1}{4}I_2 + \frac{1}{2}\sum_{j=1}^{3} n_j\sigma_j + \frac{1}{4}\sum_{j=1}^{3}\sum_{k=1}^{3} n_j n_k \sigma_j \sigma_k$$

simplifies to

$$\Pi^2(\mathbf{n}) = \frac{1}{4}I_2 + \frac{1}{2}\sum_{j=1}^{3} n_j\sigma_j + \frac{1}{4}\sum_{j=1}^{3} n_j^2 I_2 .$$

Using $n_1^2 + n_2^2 + n_3^2 = 1$ we obtain $\Pi^2(\mathbf{n}) = \Pi(\mathbf{n})$.

(ii) We find

$$\Pi(\mathbf{n}) \begin{pmatrix} e^{i\phi}\cos\theta \\ \sin\theta \end{pmatrix} = \frac{1}{2}\begin{pmatrix} (1 + n_3)e^{i\phi}\cos\theta + (n_1 - in_2)\sin\theta \\ (n_1 + in_2)e^{i\phi}\cos\theta + (1 - n_3)\sin\theta \end{pmatrix} .$$

Problem 15.   Let

$$\mathbf{x} = \begin{pmatrix} e^{i\phi}\cos\theta \\ \sin\theta \end{pmatrix}$$

where $\phi, \theta \in \mathbf{R}$.

(i) Find the matrix $\rho := \mathbf{x}\mathbf{x}^*$.

(ii) Find $\mathrm{tr}\rho$.
(iii) Find ρ^2.

Solution 15.   (i) Since $\mathbf{x}^* = (e^{-i\phi}\cos\theta,\ \sin\theta)$ we obtain the 2×2 matrix

$$\rho = \mathbf{x}\mathbf{x}^* = \begin{pmatrix} \cos^2\theta & e^{i\phi}\sin\theta\cos\theta \\ e^{-i\phi}\sin\theta\cos\theta & \sin^2\theta \end{pmatrix}.$$

(ii) Since $\cos^2\theta + \sin^2\theta = 1$ we obtain from (i) that $\mathrm{tr}\rho = 1$.
(iii) We have

$$\rho^2 = (\mathbf{x}\mathbf{x}^*)(\mathbf{x}\mathbf{x}^*) = \mathbf{x}(\mathbf{x}^*\mathbf{x})\mathbf{x}^* = \mathbf{x}\mathbf{x}^* = \rho$$

since $\mathbf{x}^*\mathbf{x} = 1$.

Problem 16.   Consider the vector space $\mathbf{R}^4$. Find all pairwise orthogonal vectors (column vectors) $\mathbf{x}_1, \dots, \mathbf{x}_p$, where the entries of the column vectors can only be $+1$ or -1. Calculate the matrix

$$\sum_{j=1}^{p} \mathbf{x}_j\mathbf{x}_j^T$$

and find the eigenvalues and eigenvectors of this matrix.

Solution 16.   The number of vectors p cannot exceed 4 since that would imply $\dim(\mathbf{R}^4) > 4$. A solution is

$$\mathbf{x}_1 = \begin{pmatrix} 1 \\ 1 \\ 1 \\ 1 \end{pmatrix}, \quad \mathbf{x}_2 = \begin{pmatrix} 1 \\ -1 \\ 1 \\ -1 \end{pmatrix}, \quad \mathbf{x}_3 = \begin{pmatrix} 1 \\ -1 \\ -1 \\ 1 \end{pmatrix}, \quad \mathbf{x}_4 = \begin{pmatrix} 1 \\ 1 \\ -1 \\ -1 \end{pmatrix}.$$

Thus

$$\sum_{j=1}^{4} \mathbf{x}_j\mathbf{x}_j^T = \begin{pmatrix} 1 & 1 & 1 & 1 \\ 1 & 1 & 1 & 1 \\ 1 & 1 & 1 & 1 \\ 1 & 1 & 1 & 1 \end{pmatrix} + \begin{pmatrix} 1 & -1 & 1 & -1 \\ -1 & 1 & -1 & 1 \\ 1 & -1 & 1 & -1 \\ -1 & 1 & -1 & 1 \end{pmatrix}$$

$$+ \begin{pmatrix} 1 & -1 & -1 & 1 \\ -1 & 1 & 1 & -1 \\ -1 & 1 & 1 & -1 \\ 1 & -1 & -1 & 1 \end{pmatrix} + \begin{pmatrix} 1 & 1 & -1 & -1 \\ 1 & 1 & -1 & -1 \\ -1 & -1 & 1 & 1 \\ -1 & -1 & 1 & 1 \end{pmatrix}$$

$$= \begin{pmatrix} 4 & 0 & 0 & 0 \\ 0 & 4 & 0 & 0 \\ 0 & 0 & 4 & 0 \\ 0 & 0 & 0 & 4 \end{pmatrix}.$$

The eigenvalue is 4 with multiplicity 4. The eigenvectors are all $\mathbf{x} \in \mathbf{R}^4$ with $\mathbf{x} \neq \mathbf{0}$. Another solution is given by

$$\mathbf{x}_1 = \begin{pmatrix} 1 \\ 1 \\ 1 \\ -1 \end{pmatrix}, \quad \mathbf{x}_2 = \begin{pmatrix} 1 \\ 1 \\ -1 \\ 1 \end{pmatrix}, \quad \mathbf{x}_3 = \begin{pmatrix} 1 \\ -1 \\ 1 \\ 1 \end{pmatrix}, \quad \mathbf{x}_4 = \begin{pmatrix} -1 \\ 1 \\ 1 \\ 1 \end{pmatrix}.$$

Problem 17. Let

$$A = \begin{pmatrix} 2 & 2 & -2 \\ 2 & 2 & -2 \\ -2 & -2 & 6 \end{pmatrix}.$$

(i) Let X be an $m \times n$ matrix. The *column rank* of X is the maximum number of linearly independent columns. The *row rank* is the maximum number of linearly independent rows. The row rank and the column rank of X are equal (called the *rank* of X). Find the rank of A and denote it by k.

(ii) Locate a $k \times k$ submatrix of A having rank k.

(iii) Find 3×3 permutation matrices P and Q such that in the matrix PAQ the submatrix from (ii) is in the upper left portion of A.

Solution 17. (i) The vectors in the first two columns are linearly dependent. Thus the rank of A is 2.

(ii) A 2×2 submatrix having rank 2 is

$$B = \begin{pmatrix} 2 & -2 \\ -2 & 6 \end{pmatrix}.$$

(iii) Let

$$P = Q = \begin{pmatrix} 1 & 0 & 0 \\ 0 & 0 & 1 \\ 0 & 1 & 0 \end{pmatrix}.$$

Then

$$PAQ = \begin{pmatrix} 2 & -2 & 2 \\ -2 & 6 & -2 \\ 2 & -2 & 2 \end{pmatrix}.$$

Problem 18. Find 2×2 matrices A, B such that $AB = 0_n$ and $BA \neq 0_n$.

Solution 18. An example is

$$A = \begin{pmatrix} 0 & 1 \\ 0 & 0 \end{pmatrix}, \quad B = \begin{pmatrix} 1 & 0 \\ 0 & 0 \end{pmatrix}.$$

Problem 19. Let A be an $m \times n$ matrix and B be a $p \times q$ matrix. Then the *direct sum* of A and B, denoted by $A \oplus B$, is the $(m + p) \times (n + q)$ matrix defined by

$$A \oplus B := \begin{pmatrix} A & 0 \\ 0 & B \end{pmatrix}.$$

Let A_1, A_2 be $m \times m$ matrices and B_1, B_2 be $n \times n$ matrices. Calculate

$$(A_1 \oplus B_1)(A_2 \oplus B_2).$$

Solution 19. We find

$$(A_1 \oplus B_1)(A_2 \oplus B_2) = (A_1 A_2) \oplus (B_1 B_2).$$

Problem 20. Let A be an $n \times n$ matrix over **R**. Find all matrices that satisfy the equation

$$A^T A = 0_n.$$

Solution 20. From $A^T A = 0_n$ we find

$$\sum_{i=1}^{n} a_{ij} a_{ij} = 0$$

where $j = 1, 2, \ldots, n$. Thus A must be the zero matrix.

Problem 21. Let π be a permutation on $\{1, 2, \ldots, n\}$. The matrix P_π for which $p_{i*} = e_{\pi(i)*}$ is called the *permutation matrix* associated with π, where p_{i*} is the ith row of P_π and $e_{ij} = 1$ if $i = j$ and 0 otherwise. Let $\pi = (3\ 2\ 4\ 1)$. Find P_π.

Solution 21. P_π is obtained from the $n \times n$ identity matrix I_n by applying π to the rows of I_n. We find

$$P_\pi = \begin{pmatrix} 0 & 0 & 1 & 0 \\ 0 & 1 & 0 & 0 \\ 0 & 0 & 0 & 1 \\ 1 & 0 & 0 & 0 \end{pmatrix}$$

and

$$p_{1*} = e_{3*} = e_{\pi(1)*}$$
$$p_{2*} = e_{2*} = e_{\pi(2)*}$$
$$p_{3*} = e_{4*} = e_{\pi(3)*}$$
$$p_{4*} = e_{1*} = e_{\pi(4)*}.$$

Problem 22. A matrix A for which $A^p = 0_n$, where p is a positive integer, is called *nilpotent*. If p is the least positive integer for which $A^p = 0_n$ then A is said to be nilpotent of index p. Find all 2×2 matrices over the real numbers which are nilpotent with $p = 2$, i.e. $A^2 = 0_2$.

Solution 22. Let
$$A = \begin{pmatrix} a_{11} & a_{12} \\ a_{21} & a_{22} \end{pmatrix} .$$
Then from $A^2 = 0_2$ we obtain the four equations
$$a_{11}^2 + a_{12}a_{21} = 0, \quad a_{12}(a_{11}+a_{22}) = 0, \quad a_{21}(a_{11}+a_{22}) = 0, \quad a_{12}a_{21}+a_{22}^2 = 0 .$$

Thus we have to consider the cases $a_{11} + a_{22} \neq 0$ and $a_{11} + a_{22} = 0$. If $a_{11} + a_{22} \neq 0$, then $a_{12} = a_{21} = 0$ and therefore $a_{11} = a_{22} = 0$. Thus we have the 2×2 zero matrix. If $a_{11}+a_{22} = 0$ we have $a_{11} = -a_{22}$ and $a_{11} \neq 0$, otherwise we would find the zero matrix again. Thus $a_{12}a_{21} = -a_{11}^2 = -a_{22}^2$ and for this case we find the solution
$$A = \begin{pmatrix} a_{11} & a_{12} \\ -a_{11}^2/a_{12} & -a_{11} \end{pmatrix} .$$

Problem 23. A square matrix is called *idempotent* if $A^2 = A$. Find all 2×2 matrices over the real numbers which are idempotent and $a_{ij} \neq 0$ for $i, j = 1, 2$.

Solution 23. Let
$$A = \begin{pmatrix} a_{11} & a_{12} \\ a_{21} & a_{22} \end{pmatrix} .$$
Then from $A^2 = A$ we obtain
$$a_{11}^2 + a_{12}a_{21} = a_{11}, \quad a_{12}(a_{11} + a_{22}) = a_{12},$$
$$a_{21}(a_{11} + a_{22}) = a_{21}, \quad a_{12}a_{21} + a_{22}^2 = a_{22} .$$
Since $a_{ij} \neq 0$ we obtain $a_{11} + a_{22} = 1$ and
$$a_{21} = \frac{1}{a_{12}}(a_{11} - a_{11}^2) .$$

Thus the matrix is
$$A = \begin{pmatrix} a_{11} & a_{12} \\ (a_{11} - a_{11}^2)/a_{12} & 1 - a_{11} \end{pmatrix}$$
with a_{11} and a_{12} arbitrary and nonzero.

Problem 24. A square matrix A such that $A^2 = I_n$ is called *involutory*. Find all 2×2 matrices over the real numbers which are involutory. Assume that $a_{ij} \neq 0$ for $i, j = 1, 2$.

Solution 24. Let

$$A = \begin{pmatrix} a_{11} & a_{12} \\ a_{21} & a_{22} \end{pmatrix}.$$

Then from $A^2 = I_2$ we obtain

$$a_{11}^2 + a_{12}a_{21} = 1, \quad a_{12}(a_{11}+a_{22}) = 0, \quad a_{21}(a_{11}+a_{22}) = 0, \quad a_{12}a_{21}+a_{22}^2 = 1.$$

Since $a_{ij} \neq 0$ we have $a_{11} + a_{22} = 0$ and

$$a_{21} = (1 - a_{11}^2)/a_{12}.$$

Then the matrix is given by

$$A = \begin{pmatrix} a_{11} & a_{12} \\ (1 - a_{11}^2)/a_{12} & -a_{11} \end{pmatrix}.$$

Problem 25. Show that an $n \times n$ matrix A is involutory if and only if $(I_n - A)(I_n + A) = 0_n$.

Solution 25. Suppose that $(I_n - A)(I_n + A) = 0_n$. Then

$$(I_n - A)(I_n + A) = I_n - A^2 = 0_n.$$

Thus $A^2 = I_n$ and A is involutory. Suppose that A is involutory. Then $A^2 = I_n$ and

$$0_n = I_n - A^2 = (I_n - A)(I_n + A).$$

Problem 26. Let A be an $n \times n$ symmetric matrix over $\mathbf{R}$. Let P be an arbitrary $n \times n$ matrix over $\mathbf{R}$. Show that $P^T A P$ is symmetric.

Solution 26. Using that $A^T = A$ and $(P^T)^T = P$ we have

$$(P^T A P)^T = P^T A^T (P^T)^T = P^T A P.$$

Thus $P^T A P$ is symmetric.

Problem 27. Let A be an $n \times n$ skew-symmetric matrix over $\mathbf{R}$, i.e. $A^T = -A$. Let P be an arbitrary $n \times n$ matrix over $\mathbf{R}$. Show that $P^T A P$ is skew-symmetric.

Solution 27. Using $A^T = -A$ and $(P^T)^T = P$ we have

$$(P^T A P)^T = P^T A^T (P^T)^T = -P^T A P.$$

Thus $P^T A P$ is skew-symmetric.

Problem 28. Let A be an $m \times n$ matrix. The *column rank* of A is the maximum number of linearly independent columns. The *row rank* is the maximum number of linearly independent rows. The row rank and the column rank of A are equal (called the rank of A). Find the rank of the 4×4 matrix

$$A = \begin{pmatrix} 1 & 2 & 3 & 4 \\ 5 & 6 & 7 & 8 \\ 9 & 10 & 11 & 12 \\ 13 & 14 & 15 & 16 \end{pmatrix}.$$

Solution 28. The *elementary transformations* do not change the rank of a matrix. We subtract the third column from the fourth column, the second column from the third column and the first column from the second column, i.e.

$$\begin{pmatrix} 1 & 2 & 3 & 1 \\ 5 & 6 & 7 & 1 \\ 9 & 10 & 11 & 1 \\ 13 & 14 & 15 & 1 \end{pmatrix} \sim \begin{pmatrix} 1 & 2 & 1 & 1 \\ 5 & 6 & 1 & 1 \\ 9 & 10 & 1 & 1 \\ 13 & 14 & 1 & 1 \end{pmatrix} \sim \begin{pmatrix} 1 & 1 & 1 & 1 \\ 5 & 1 & 1 & 1 \\ 9 & 1 & 1 & 1 \\ 13 & 1 & 1 & 1 \end{pmatrix}.$$

From the last matrix we see (three columns are the same) that the rank of A is 2. It follows that two eigenvalues must be 0.

Problem 29. Let A be an invertible $n \times n$ matrix over $\mathbf{C}$ and B be an $n \times n$ matrix over $\mathbf{C}$. We define the $n \times n$ matrix

$$D := A^{-1} B A.$$

Calculate D^n, where $n = 2, 3, \ldots$.

Solution 29. Since $A A^{-1} = I_n$ we obtain

$$D^n = A^{-1} B^n A.$$

Problem 30. A *Cartan matrix* A is a square matrix whose elements a_{ij} satisfy the following conditions:
1. a_{ij} is an integer, one of $\{-3, -2, -1, 0, 2\}$
2. $a_{jj} = 2$ for all diagonal elements of A

3. $a_{ij} \leq 0$ off of the diagonal
4. $a_{ij} = 0$ iff $a_{ji} = 0$
5. There exists an invertible diagonal matrix D such that DAD^{-1} gives a symmetric and positive definite quadratic form.

Give a 2×2 non-diagonal Cartan matrix.

Solution 30. An example is

$$A = \begin{pmatrix} 2 & -1 \\ -1 & 2 \end{pmatrix}.$$

The first four conditions are obvious for the matrix A. The last condition can be seen from

$$\mathbf{x}^T A \mathbf{x} = \begin{pmatrix} x_1 & x_2 \end{pmatrix} \begin{pmatrix} 2 & -1 \\ -1 & 2 \end{pmatrix} \begin{pmatrix} x_1 \\ x_2 \end{pmatrix} = 2(x_1^2 - x_1 x_2 + x_2^2) \geq 0$$

for $\mathbf{x} \neq 0$. That the symmetric matrix A is positive definite can also be seen from the eigenvalues of A which are 3 and 1.

Problem 31. Let A, B, C, D be $n \times n$ matrices over **R**. Assume that AB^T and CD^T are symmetric and $AD^T - BC^T = I_n$, where ^{T} denotes transpose. Show that

$$A^T D - C^T B = I_n.$$

Solution 31. From the assumptions we have

$$AB^T = (AB^T)^T = BA^T$$
$$CD^T = (CD^T)^T = DC^T$$
$$AD^T - BC^T = I_n.$$

Taking the transpose of the third equation we have

$$DA^T - CB^T = I_n.$$

These four equations can be written in the form of block matrices in the identity

$$\begin{pmatrix} A & B \\ C & D \end{pmatrix} \begin{pmatrix} D^T & -B^T \\ -C^T & A^T \end{pmatrix} = \begin{pmatrix} I_n & 0_n \\ 0_n & I_n \end{pmatrix}.$$

Thus the matrices are $(2n) \times (2n)$ marices. If X, Y are $m \times m$ matrices with $XY = I_m$, the identity matrix, then $Y = X^{-1}$ and $YX = I_m$ too. Applying this to our matrix equation with $m = 2n$ we obtain

$$\begin{pmatrix} D^T & -B^T \\ -C^T & A^T \end{pmatrix} \begin{pmatrix} A & B \\ C & D \end{pmatrix} = \begin{pmatrix} I_n & 0_n \\ 0_n & I_n \end{pmatrix}.$$

Equating the lower right blocks shows that $-C^T B + A^T D = I_n$.

Problem 32. Let n be a positive integer. Let A_n be the $(2n+1) \times (2n+1)$ skew-symmetric matrix for which each entry in the first n subdiagonals below the main diagonal is 1 and each of the remaining entries below the main diagonal is -1. Give A_1 and A_2. Find the rank of A_n.

Solution 32. We have

$$A_1 = \begin{pmatrix} 0 & -1 & 1 \\ 1 & 0 & -1 \\ -1 & 1 & 0 \end{pmatrix}$$

$$A_2 = \begin{pmatrix} 0 & -1 & -1 & 1 & 1 \\ 1 & 0 & -1 & -1 & 1 \\ 1 & 1 & 0 & -1 & -1 \\ -1 & 1 & 1 & 0 & -1 \\ -1 & -1 & 1 & 1 & 0 \end{pmatrix}.$$

We use induction on n to prove that $\text{rank}(A_n) = 2n$. The rank of A_1 is 2 since the first vector in the matrix A_1 is linear combination of the second and third vectors

$$\begin{pmatrix} 1 \\ -1 \\ 0 \end{pmatrix} = -\begin{pmatrix} 0 \\ 1 \\ -1 \end{pmatrix} - \begin{pmatrix} -1 \\ 0 \\ 1 \end{pmatrix}$$

and the second and third vectors are linearly independent. Suppose $n \geq 2$ and that $\text{rank}(A_{n-1}) = 2(n-1)$ is known. Adding multiples of the first two rows of A_n to the other rows transforms A_n to a matrix of the form

$$\begin{pmatrix} 0 & -1 & \\ 1 & 0 & * \\ 0 & & -A_{n-1} \end{pmatrix}$$

in which 0 and $*$ represent blocks of size $(2n-1) \times 2$ and $2 \times (2n-1)$, repectively. Thus $\text{rank}(A_n) = 2 + \text{rank}(A_{n-1}) = 2 + 2(n-1) = 2n$.

Problem 33. A vector $\mathbf{u} = (u_1, u_2, \ldots, u_n)$ is called a *probability vector* if the components are nonnegative and their sum is 1. Is the vector

$$\mathbf{u} = (1/2,\ 0,\ 1/4,\ 1/4)$$

a probability vector? Can the vector

$$\mathbf{v} = (2, 3, 5, 1, 0)$$

be "normalized" so that we obtain a probability vector?

Solution 33.   Since all the components are nonnegative and the sum of entries is 1 we find that **u** is a probability vector. All the entries in **v** are nonnegative but the sum is 11. Thus we can construct the probability vector

$$\tilde{\mathbf{v}} = \frac{1}{11}(2, 3, 5, 1, 0).$$

Problem 34.   An $n \times n$ matrix $P = (p_{ij})$ is called a *stochastic matrix* if each of its rows is a probability vector, i.e., if each entry of P is nonnegative and the sum of the entries in each row is 1. Let A and B be two stochastic $n \times n$ matrices. Is the matrix product AB also a stochastic matrix?

Solution 34.   From matrix multiplication we have for the (ij) entry of the product

$$(AB)_{ij} = \sum_{k=1}^{n} a_{ik}b_{kj}.$$

It follows that

$$\sum_{j=1}^{n}(AB)_{ij} = \sum_{j=1}^{n}\sum_{k=1}^{n} a_{ik}b_{kj} = \sum_{k=1}^{n} a_{ik} \sum_{j=1}^{n} b_{kj} = 1$$

since

$$\sum_{k=1}^{n} a_{ik} = 1, \qquad \sum_{j=1}^{n} b_{kj} = 1.$$

Problem 35.   The *numerical range*, also known as the *field of values*, of an $n \times n$ matrix A over the complex numbers, is defined as

$$F(A) := \{\, \mathbf{z}^* A \mathbf{z} \; : \; \|\mathbf{z}\| = 1, \; \mathbf{z} \in \mathbf{C}^n \,\}.$$

Find the numerical range for the 2×2 matrix

$$B = \begin{pmatrix} 1 & 0 \\ 0 & 0 \end{pmatrix}.$$

Find the numerical range for the 2×2 matrix

$$C = \begin{pmatrix} 0 & 0 \\ 1 & 1 \end{pmatrix}.$$

The *Toeplitz-Hausdorff convexity theorem* tells us that the numerical range of a square matrix is a convex compact subset of the complex plane.

Solution 35. Let

$$A = \begin{pmatrix} a_{11} & a_{12} \\ a_{21} & a_{22} \end{pmatrix}$$

and

$$\mathbf{z} = \begin{pmatrix} e^{i\phi}\cos\theta \\ e^{i\chi}\sin\theta \end{pmatrix}, \qquad \phi, \chi, \theta \in \mathbf{R}.$$

Therefore $\mathbf{z}$ is an arbitrary complex number of length 1, i.e., $\|\mathbf{z}\| = 1$. Then

$$\mathbf{z}^* A\mathbf{z} = \begin{pmatrix} e^{-i\phi}\cos\theta & e^{-i\chi}\sin\theta \end{pmatrix} \begin{pmatrix} a_{11} & a_{12} \\ a_{21} & a_{22} \end{pmatrix} \begin{pmatrix} e^{i\phi}\cos\theta \\ e^{i\chi}\sin\theta \end{pmatrix}$$

$$= a_{11}\cos^2\theta + (a_{12}e^{i(\chi-\phi)} + a_{21}e^{i(\phi-\chi)})\sin\theta\cos\theta + a_{22}\sin^2\theta.$$

Thus for the matrix B we have $\mathbf{z}^T B\mathbf{z} = \cos^2\theta$, where $\cos^2\theta \in [0,1]$ for all $\theta \in \mathbf{R}$. For the matrix C we have

$$\mathbf{z}^* C\mathbf{z} = e^{i(\phi-\chi)}\sin\theta\cos\theta + \sin^2\theta.$$

Thus the numerical range $F(C)$ is the closed elliptical disc in the complex plane with foci at $(0,0)$ and $(1,0)$, minor axis 1, and major axis $\sqrt{2}$.

Problem 36. Let A be an $n \times n$ matrix over $\mathbf{C}$. The *field of values* of A is defined as the set

$$F(A) := \{ \mathbf{z}^* A\mathbf{z} : \mathbf{z} \in \mathbf{C}^n, \ \mathbf{z}^*\mathbf{z} = 1 \}.$$

Let $\alpha \in \mathbf{R}$ and

$$A = \begin{pmatrix} \alpha & 1 & 0 & 0 & 0 & 0 & 0 & 0 & 0 \\ 1 & \alpha & 1 & 0 & 0 & 0 & 0 & 0 & 0 \\ 0 & 1 & \alpha & 1 & 0 & 0 & 0 & 0 & 0 \\ 0 & 0 & 1 & \alpha & 1 & 0 & 0 & 0 & 0 \\ 0 & 0 & 0 & 1 & \alpha & 1 & 0 & 0 & 0 \\ 0 & 0 & 0 & 0 & 1 & \alpha & 1 & 0 & 0 \\ 0 & 0 & 0 & 0 & 0 & 1 & \alpha & 1 & 0 \\ 0 & 0 & 0 & 0 & 0 & 0 & 1 & \alpha & 1 \\ 0 & 0 & 0 & 0 & 0 & 0 & 0 & 1 & \alpha \end{pmatrix}$$

(i) Show that the set $F(A)$ lies on the real axis.
(ii) Show that

$$|\mathbf{z}^* A\mathbf{z}| \le \alpha + 16.$$

Solution 36. (i) Since

$$\mathbf{z}^* = (\overline{z_1}, \overline{z_2}, \overline{z_3}, \overline{z_4}, \overline{z_5}, \overline{z_6}, \overline{z_7}, \overline{z_8}, \overline{z_9})$$

with $z^*z = 1$ and applying matrix multiplication we obtain

$$z^*Az = \alpha + \sum_{j=1}^{8}(\overline{z_j}z_{j+1} + \overline{z_{j+1}}z_j).$$

Let $z_j = r_je^{i\theta_j}$. Then $\overline{z_j} = r_je^{-i\theta_j}$. Let $j \neq k$. Then

$$\overline{z_j}z_k + z_j\overline{z_k} = 2r_jr_k\cos(\theta_j - \theta_k)$$

with $0 \leq r_j \leq 1$. Thus we have

$$z^*Az = \alpha + 2\sum_{j=1}^{8}r_jr_{j+1}\cos(\theta_{j+1} - \theta_j).$$

Thus the set $F(A)$ lies on the real axis.
(ii) Since $0 \leq r_j \leq 1$ and $|\cos\theta| \leq 1$ we obtain

$$|z^*Az| \leq \alpha + 16.$$

Problem 37.   Let A be an $n \times n$ matrix over $\mathbf{C}$ and $F(A)$ the field of values. Let U be an $n \times n$ unitary matrix.
(i) Show that

$$F(U^*AU) = F(A).$$

(ii) Apply the theorem to the two matrices

$$A_1 = \begin{pmatrix} 0 & 1 \\ 1 & 0 \end{pmatrix}, \qquad A_2 = \begin{pmatrix} 1 & 0 \\ 0 & -1 \end{pmatrix}$$

which are unitarily equivalent.

Solution 37.   (i) Since a unitary matrix leaves invariant the surface of the Euclidean unit ball, the complex numbers that comprise the sets $F(U^*AU)$ and $F(A)$ are the same. If $z \in \mathbf{C}^n$ and $z^*z = 1$, we have

$$z^*(U^*AU)z = w^*Aw \in F(A)$$

where $w = Uz$, so that $w^*w = z^*U^*Uz = z^*z = 1$. Thus $F(U^*AU) \subset F(A)$. The reverse containment is obtained similarly.
(ii) For A_1 we have

$$z^*A_1z = \overline{z_1}z_2 + \overline{z_2}z_1 = 2r_1r_2\cos(\theta_2 - \theta_1)$$

with the constraints $0 \leq r_1, r_2 \leq 1$ and $r_1^2 + r_2^2 = 1$. For A_2 we find

$$z^*A_2z = \overline{z_1}z_1 - \overline{z_2}z_2 = r_1^2 - r_2^2$$

with the constraints $0 \leq r_1, r_2 \leq 1$ and $r_1^2 + r_2^2 = 1$. Both define the same set, namely the interval $[-1, 1]$.

Problem 38.   Can one find a unitary matrix U such that

$$U^* \begin{pmatrix} 0 & c \\ d & 0 \end{pmatrix} U = \begin{pmatrix} 0 & ce^{i\theta} \\ de^{-i\theta} & 0 \end{pmatrix}$$

where $c, d \in \mathbf{C}$ and $\theta \in \mathbf{R}$?

Solution 38.   We find

$$U = \begin{pmatrix} 1 & 0 \\ 0 & e^{i\theta} \end{pmatrix} .$$

Problem 39.   An $n^2 \times n$ matrix J is called a *selection matrix* such that J^T is the $n \times n^2$ matrix

$$[E_{11} \ E_{22} \ \ldots \ E_{nn}]$$

where E_{ii} is the $n \times n$ matrix of zeros except for a 1 in the (i, i)th position.
(i) Find J for $n = 2$ and calculate $J^T J$.
(ii) Calculate $J^T J$ for arbitrary n.

Solution 39.   (i) We have

$$J^T = \begin{pmatrix} 1 & 0 & 0 & 0 \\ 0 & 0 & 0 & 1 \end{pmatrix} .$$

Thus

$$J = \begin{pmatrix} 1 & 0 \\ 0 & 0 \\ 0 & 0 \\ 0 & 1 \end{pmatrix} .$$

Therefore $J^T J = I_2$.
(ii) For the general case we find $J^T J = I_n$.

Problem 40.   Let A and B be $m \times n$ matrices. The *Hadamard product* $A \circ B$ is defined as the $m \times n$ matrix

$$A \bullet B := (a_{ij} b_{ij}) .$$

(i) Let

$$A = \begin{pmatrix} 0 & 1 \\ 1 & 0 \end{pmatrix}, \qquad B = \begin{pmatrix} 3 & 4 \\ 7 & 1 \end{pmatrix} .$$

Calculate $A \bullet B$.

(ii) Let C, D be $m \times n$ matrices. Show that

$$\text{rank}(A \bullet B) \leq (\text{rank}A)(\text{rank}B).$$

Solution 40.   (i) Entrywise multiplication of the two matrices yields

$$A \bullet B = \begin{pmatrix} 0 & 4 \\ 7 & 0 \end{pmatrix}.$$

(ii) Any matrix of rank r can be written as a sum of r rank one matrices, each of which is an outer product of two vectors (column vector times row vector). Thus, if $\text{rank}A = r_1$ and $\text{rank}B = r_2$, we have

$$A = \sum_{i=1}^{r_1} \mathbf{x}_i \mathbf{y}_i^*, \qquad B = \sum_{j=1}^{r_2} \mathbf{u}_j \mathbf{v}_j^*$$

where $\mathbf{x}_i, \mathbf{u}_j \in \mathbf{C}^m$, $\mathbf{y}_i, \mathbf{v}_j \in \mathbf{C}^n$, $i = 1, \dots, r_1$ and $j = 1, \dots, r_2$. Then

$$A \bullet B = \sum_{i=1}^{r_1} \sum_{j=1}^{r_2} (\mathbf{x}_i \bullet \mathbf{u}_j)(\mathbf{y}_i \bullet \mathbf{v}_j)^*.$$

This shows that $A \bullet B$ is a sum of at most $r_1 r_2$ rank one matrices. Thus $\text{rank}(A \bullet B) \leq r_1 r_2 = (\text{rank}A)(\text{rank}B)$.

Problem 41.   Consider a symmetric matrix A over $\mathbf{R}$

$$A = \begin{pmatrix} a_{11} & a_{12} & a_{13} & a_{14} \\ a_{12} & a_{22} & a_{23} & a_{24} \\ a_{13} & a_{23} & a_{33} & a_{34} \\ a_{14} & a_{24} & a_{34} & a_{44} \end{pmatrix}$$

and the orthonormal basis (so-called *Bell basis*)

$$\mathbf{x}^+ = \frac{1}{\sqrt{2}} \begin{pmatrix} 1 \\ 0 \\ 0 \\ 1 \end{pmatrix}, \qquad \mathbf{x}^- = \frac{1}{\sqrt{2}} \begin{pmatrix} 1 \\ 0 \\ 0 \\ -1 \end{pmatrix}$$

$$\mathbf{y}^+ = \frac{1}{\sqrt{2}} \begin{pmatrix} 0 \\ 1 \\ 1 \\ 0 \end{pmatrix}, \qquad \mathbf{y}^- = \frac{1}{\sqrt{2}} \begin{pmatrix} 0 \\ 1 \\ -1 \\ 0 \end{pmatrix}.$$

The Bell basis forms an orthonormal basis in $\mathbf{R}^4$. Let $\tilde{A}$ denote the matrix A in the Bell basis. What is the condition on the entries a_{ij} such that the matrix A is diagonal in the Bell basis?

Solution 41. Obviously we have

$$\tilde{a}_{ij} = \tilde{a}_{ji},$$

i.e., the matrix $\tilde{A}$ is also symmetric. Straightforward calculation yields

$$\tilde{a}_{11} = (\mathbf{x}^+)^T A \mathbf{x}^+ = \frac{1}{2}(a_{11} + 2a_{14} + a_{44})$$

$$\tilde{a}_{12} = (\mathbf{x}^+)^T A \mathbf{x}^- = \frac{1}{2}(a_{11} - a_{44})$$

$$\tilde{a}_{13} = (\mathbf{x}^+)^T A \mathbf{y}^+ = \frac{1}{2}(a_{12} + a_{13} + a_{24} + a_{34})$$

$$\tilde{a}_{14} = (\mathbf{x}^+)^T A \mathbf{y}^- = \frac{1}{2}(a_{12} - a_{13} + a_{24} - a_{34})$$

$$\tilde{a}_{22} = (\mathbf{x}^-)^T A \mathbf{x}^- = \frac{1}{2}(a_{11} - 2a_{14} + a_{44})$$

$$\tilde{a}_{23} = (\mathbf{x}^-)^T A \mathbf{y}^+ = \frac{1}{2}(a_{12} + a_{13} - a_{24} - a_{34})$$

$$\tilde{a}_{24} = (\mathbf{x}^-)^T A \mathbf{y}^- = \frac{1}{2}(a_{12} - a_{13} - a_{24} + a_{34})$$

$$\tilde{a}_{33} = (\mathbf{y}^+)^T A \mathbf{y}^+ = \frac{1}{2}(a_{22} + 2a_{23} + a_{33})$$

$$\tilde{a}_{34} = (\mathbf{y}^+)^T A \mathbf{y}^- = \frac{1}{2}(a_{22} - a_{33})$$

$$\tilde{a}_{44} = (\mathbf{y}^-)^T A \mathbf{y}^- = \frac{1}{2}(a_{22} - 2a_{23} + a_{33}).$$

The condition that the matrix $\tilde{A}$ should be diagonal leads to

$$a_{11} - a_{44} = 0, \qquad a_{22} - a_{33} = 0$$

and

$$a_{12} = a_{13} = a_{24} = a_{34} = 0$$

with the entries a_{14} and a_{23} arbitrary. Thus the matrix A has the form

$$A = \begin{pmatrix} a_{11} & 0 & 0 & a_{14} \\ 0 & a_{22} & a_{23} & 0 \\ 0 & a_{23} & a_{22} & 0 \\ a_{14} & 0 & 0 & a_{11} \end{pmatrix}.$$

Problem 42. Let A be an $m \times n$ matrix over $\mathbf{C}$. The *Moore-Penrose pseudoinverse matrix* A^+ is the unique $n \times m$ matrix which satisfies

$$AA^+A = A$$
$$A^+AA^+ = A^+$$
$$(AA^+)^* = AA^+$$
$$(A^+A)^* = A^+A.$$

We also have that

$$\mathbf{x} = A^+ \mathbf{b} \tag{1}$$

is the shortest length least square solution to the problem

$$A\mathbf{x} = \mathbf{b}. \tag{2}$$

(i) Show that if $(A^*A)^{-1}$ exists, then $A^+ = (A^*A)^{-1}A^*$.
(ii) Let

$$A = \begin{pmatrix} 1 & 3 \\ 2 & 4 \\ 3 & 5 \end{pmatrix}.$$

Find the Moore-Penrose matrix inverse A^+ of A.

Solution 42.   (i) Suppose that $(A^*A)^{-1}$ exists we have

$$A\mathbf{x} = \mathbf{b}$$
$$A^*A\mathbf{x} = A^*\mathbf{b}$$
$$\mathbf{x} = (A^*A)^{-1}A^*\mathbf{b}.$$

Using (1) we obtain

$$A^+ = (A^*A)^{-1}A^*.$$

(ii) We have

$$A^*A = \begin{pmatrix} 1 & 2 & 3 \\ 3 & 4 & 5 \end{pmatrix} \begin{pmatrix} 1 & 3 \\ 2 & 4 \\ 3 & 5 \end{pmatrix} = \begin{pmatrix} 14 & 26 \\ 26 & 50 \end{pmatrix}.$$

Since $\det(A^*A) \neq 0$ the inverse of A^*A exists and is given by

$$(A^*A)^{-1} = \frac{1}{12} \begin{pmatrix} 25 & -13 \\ -13 & 7 \end{pmatrix}.$$

Thus

$$A^+ = (A^*A)^{-1}A^* = \frac{1}{12} \begin{pmatrix} 25 & -13 \\ -13 & 7 \end{pmatrix} \begin{pmatrix} 1 & 2 & 3 \\ 3 & 4 & 5 \end{pmatrix}$$
$$= \frac{1}{12} \begin{pmatrix} -14 & -2 & 10 \\ 8 & 2 & -4 \end{pmatrix}.$$

Problem 43.   Given a signal as the column vector

$$\mathbf{x} = (3.0 \ \ 0.5 \ \ 2.0 \ \ 7.0)^T.$$

The *pyramid algorithm* (for *Haar wavelets*) is as follows: The first two entries $(3.0 \ 0.5)^T$ in the signal give an average of $(3.0 + 0.5)/2 = 1.75$ and a difference average of $(3.0 - 0.5)/2 = 1.25$. The second two entries $(2.0 \ 7.0)$ give an average of $(2.0 + 7.0)/2 = 4.5$ and a difference average of $(2.0 - 7.0)/2 = -2.5$. Thus we end up with a vector

$$(1.75 \ 1.25 \ 4.5 \ -2.5)^T .$$

Now we take the average of 1.75 and 4.5 providing $(1.75 + 4.5)/2 = 3.125$ and the difference average $(1.75 - 4.5)/2 = -1.375$. Thus we end up with the vector

$$\mathbf{y} = (3.125 \ -1.375 \ 1.25 \ -2.5)^T .$$

(i) Find a 4×4 matrix A such that

$$\mathbf{x} \equiv \begin{pmatrix} 3.0 \\ 0.5 \\ 2.0 \\ 7.0 \end{pmatrix} = A\mathbf{y} \equiv A \begin{pmatrix} 3.125 \\ -1.375 \\ 1.25 \\ -2.5 \end{pmatrix} .$$

(ii) Show that the inverse of A exists. Then find the inverse matrix of A.

Solution 43. (i) Since we can write

$$\begin{pmatrix} 3.0 \\ 0.5 \\ 2.0 \\ 7.0 \end{pmatrix} = 3.125 \begin{pmatrix} 1 \\ 1 \\ 1 \\ 1 \end{pmatrix} - 1.375 \begin{pmatrix} 1 \\ 1 \\ -1 \\ -1 \end{pmatrix} + 1.25 \begin{pmatrix} 1 \\ -1 \\ 0 \\ 0 \end{pmatrix} - 2.5 \begin{pmatrix} 0 \\ 0 \\ 1 \\ -1 \end{pmatrix}$$

we obtain the matrix

$$A = \begin{pmatrix} 1 & 1 & 1 & 0 \\ 1 & 1 & -1 & 0 \\ 1 & -1 & 0 & 1 \\ 1 & -1 & 0 & -1 \end{pmatrix} .$$

(ii) All the column vectors in the matrix A are nonzero and all the pairwise scalar products are equal to 0. Thus the column vectors form a basis (not normalized) in $\mathbf{R}^n$. Thus the matrix is invertible. The inverse matrix is given by

$$A^{-1} = \begin{pmatrix} 1/4 & 1/4 & 1/4 & 1/4 \\ 1/4 & 1/4 & -1/4 & -1/4 \\ 1/2 & -1/2 & 0 & 0 \\ 0 & 0 & 1/2 & -1/2 \end{pmatrix} .$$

Problem 44. A *Hadamard matrix* is an $n \times n$ matrix H with entries in $\{-1, +1\}$ such that any two distinct rows or columns of H have inner product 0. Construct a 4×4 Hadamard matrix starting from the column vector

$$\mathbf{x}_1 = (1\ 1\ 1\ 1)^T.$$

Solution 44. The vector

$$\mathbf{x}_2 = (1\ 1\ -1\ -1)^T$$

is perpendicular to the vector $\mathbf{x}_1$. Next the vector

$$\mathbf{x}_3 = (-1\ 1\ 1\ -1)^T$$

is perpendicular to $\mathbf{x}_1$ and $\mathbf{x}_2$. Finally the vector

$$\mathbf{x}_4 = (1\ -1\ 1\ -1)$$

is perpendicular to $\mathbf{x}_1$, $\mathbf{x}_2$ and $\mathbf{x}_3$. Thus we obtain the 4×4 Hadamard matrix

$$H = \begin{pmatrix} 1 & 1 & -1 & 1 \\ 1 & 1 & 1 & -1 \\ 1 & -1 & 1 & 1 \\ 1 & -1 & -1 & -1 \end{pmatrix}.$$

Problem 45. A *binary Hadamard matrix* is an $n \times n$ matrix M (where n is even) with entries in $\{0, 1\}$ such that any two distinct rows or columns of M have *Hamming distance* $n/2$. The Hamming distance between two vectors is the number of entries at which they differ. Find a 4×4 binary Hadamard matrix.

Solution 45. Hadamard matrices (see previous problem) are in *bijection* with binary Hadamard matrices with the mapping $1 \to 1$ and $-1 \to 0$. Thus using the result from the previous problem we obtain

$$M = \begin{pmatrix} 1 & 1 & 0 & 1 \\ 1 & 1 & 1 & 0 \\ 1 & 0 & 1 & 1 \\ 1 & 0 & 0 & 0 \end{pmatrix}.$$

Problem 46. Let $\mathbf{x}$ be a normalized column vector in $\mathbf{R}^n$, i.e. $\mathbf{x}^T\mathbf{x} = 1$. A matrix T is called a *Householder matrix* if

$$T := I_n - 2\mathbf{x}\mathbf{x}^T.$$

Calculate T^2.

Solution 46.   Since the matrix product is associative we have

$$T^2 = (I_n - 2\mathbf{x}\mathbf{x}^T)(I_n - 2\mathbf{x}\mathbf{x}^T)$$
$$= I_n - 2\mathbf{x}\mathbf{x}^T - 2\mathbf{x}\mathbf{x}^T + 4\mathbf{x}(\mathbf{x}^T\mathbf{x})\mathbf{x}^T$$
$$= I_n - 4\mathbf{x}\mathbf{x}^T + 4\mathbf{x}\mathbf{x}^T$$
$$= I_n .$$

Problem 47.   An $n \times n$ matrix P is a *projection matrix* if

$$P^* = P, \qquad P^2 = P.$$

(i) Let P_1 and P_2 be projection matrices. Is $P_1 + P_2$ a projection matrix?
(ii) Let P_1 and P_2 be projection matrices. Is $P_1 P_2$ a projection matrix?
(iii) Let P be a projection matrix. Is $I_n - P$ a projection matrix? Calculate $P(I_n - P)$.
(iv) Is

$$P = \frac{1}{3}\begin{pmatrix} 1 & 1 & 1 \\ 1 & 1 & 1 \\ 1 & 1 & 1 \end{pmatrix}$$

a projection matrix?

Solution 47.   (i) Obviously $(P_1 + P_2)^* = P_1^* + P_2^* = P_1 + P_2$. We have

$$(P_1 + P_2)^2 = P_1^2 + P_1 P_2 + P_2 P_1 + P_2^2 = P_1 + P_1 P_2 + P_2 P_1 + P_2 .$$

Thus $P_1 + P_2$ is a projection matrix only if $P_1 P_2 = 0_n$, where we used that from $P_1 P_2 = 0_n$ we can conclude that $P_2 P_1 = 0_n$. From $P_1 P_2 = 0_n$ it follows that

$$(P_1 P_2)^* = P_2^* P_1^* = P_2 P_1 = 0_n .$$

(ii) We have $(P_1 P_2)^* = P_2^* P_1^* = P_2 P_1$ and

$$(P_1 P_2)^2 = P_1 P_2 P_1 P_2 .$$

Thus we see that $P_1 P_2$ is a projection matrix if and only if $P_1 P_2 = P_2 P_1$.
(iii) We have $(I_n - P)^* = I_n^* - P^* = I_n - P$ and

$$(I_n - P)^2 = I_n - P - P + P^2 = I_n - P .$$

Thus $I_n - P$ is a projection matrix. We have

$$P(I_n - P) = P - P^2 = P - P = 0_n .$$

(iv) We find $P^* = P$ and $P^2 = P$. Thus P is a projection matrix.

Problem 48. Let

$$\mathbf{a} = \begin{pmatrix} a_1 \\ a_2 \\ a_3 \end{pmatrix}, \qquad \mathbf{b} = \begin{pmatrix} b_1 \\ b_2 \\ b_3 \end{pmatrix}$$

be vectors in $\mathbf{R}^3$. Let $\times$ denote the vector product.

(i) Show that we can find a 3×3 matrix $S(\mathbf{a})$ such that

$$\mathbf{a} \times \mathbf{b} = S(\mathbf{a})\mathbf{b}.$$

(ii) Express the *Jacobi identity*

$$\mathbf{a} \times (\mathbf{b} \times \mathbf{c}) + \mathbf{c} \times (\mathbf{a} \times \mathbf{b}) + \mathbf{b} \times (\mathbf{c} \times \mathbf{a}) = \mathbf{0}$$

using the matrices $S(\mathbf{a})$, $S(\mathbf{b})$ and $S(\mathbf{c})$.

Solution 48. (i) The *vector product* is defined as

$$\mathbf{a} \times \mathbf{b} := \begin{pmatrix} a_2 b_3 - a_3 b_2 \\ a_3 b_1 - a_1 b_3 \\ a_1 b_2 - a_2 b_1 \end{pmatrix}.$$

Thus $S(\mathbf{a})$ is the skew-symmetric matrix

$$S(\mathbf{a}) = \begin{pmatrix} 0 & -a_3 & a_2 \\ a_3 & 0 & -a_1 \\ -a_2 & a_1 & 0 \end{pmatrix}.$$

(ii) Using the result from (i) we obtain

$$\mathbf{a} \times (S(\mathbf{b})\mathbf{c}) + \mathbf{c} \times (S(\mathbf{a})\mathbf{b}) + \mathbf{b} \times (S(\mathbf{c})\mathbf{a}) = \mathbf{0}$$
$$S(\mathbf{a})(S(\mathbf{b})\mathbf{c}) + S(\mathbf{c})(S(\mathbf{a})\mathbf{b}) + S(\mathbf{b})(S(\mathbf{c})\mathbf{a}) = \mathbf{0}$$
$$(S(\mathbf{a})S(\mathbf{b}))\mathbf{c} + (S(\mathbf{c})S(\mathbf{a}))\mathbf{b} + (S(\mathbf{b})S(\mathbf{c}))\mathbf{a} = \mathbf{0}$$

where we used that matrix multiplication is associative.

Problem 49. Let s (*spin quantum number*)

$$s \in \left\{ \frac{1}{2}, 1, \frac{3}{2}, 2, \frac{5}{2}, \ldots \right\}.$$

Given a fixed s. The indices j, k run over $s, s-1, s-2, \ldots, -s+1, -s$. Consider the $(2s+1)$ unit vectors (standard basis)

$$\mathbf{e}_{s,s} = \begin{pmatrix} 1 \\ 0 \\ 0 \\ \vdots \\ 0 \end{pmatrix}, \quad \mathbf{e}_{s,s-1} = \begin{pmatrix} 0 \\ 1 \\ 0 \\ \vdots \\ 0 \end{pmatrix}, \quad \ldots, \mathbf{e}_{s,-s} = \begin{pmatrix} 0 \\ 0 \\ \vdots \\ 0 \\ 1 \end{pmatrix}.$$

Obviously the vectors have $(2s + 1)$ components. The $(2s + 1) \times (2s + 1)$ matrices s_+ and s_- are defined as

$$s_+ \mathbf{e}_{s,m} := \hbar \sqrt{(s - m)(s + m + 1)} \mathbf{e}_{s,m+1}, \quad m = s - 1, s - 2, \ldots, -s$$

$$s_- \mathbf{e}_{s,m} := \hbar \sqrt{(s + m)(s - m + 1)} \mathbf{e}_{s,m-1}, \quad m = s, s - 1, \ldots, -s + 1$$

and $s_+ \mathbf{e}_{ss} = \mathbf{0}$, $s_- \mathbf{e}_{s-s} = \mathbf{0}$, where $\hbar$ is the Planck constant.

(i) Find the matrix representation of s_+ and s_-.

(ii) The $(2s + 1) \times (2s + 1)$ matrix s_z is defined as

$$s_z \mathbf{e}_{s,m} := m\hbar \mathbf{e}_{s,m}, \quad m = s, s - 1, \ldots, -s.$$

Let

$$\mathbf{s} := (s_x, s_y, s_z)$$

where $s_+ = \frac{1}{2}(s_x + is_y)$ and $s_- = \frac{1}{2}(s_x - is_y)$. Find the $(2s + 1) \times (2s + 1)$ matrix

$$\mathbf{s}^2 := s_x^2 + s_y^2 + s_z^2.$$

(iii) Calculate the expectation values

$$\mathbf{e}_{s,s}^* s_+ \mathbf{e}_{s,s}, \quad \mathbf{e}_{s,s}^* s_- \mathbf{e}_{s,s}, \quad \mathbf{e}_{s,s}^* s_z \mathbf{e}_{s,s}.$$

Solution 49. (i) We have

$$(s_+)_{jk} = (s_-)_{kj} = \hbar \sqrt{(s - k)(s + k + 1)} \delta_{j,k+1}$$
$$= \hbar \sqrt{(s + j)(s - j + 1)} \delta_{j,k+1}$$

$$(s_-)_{jk} = (s_+)_{kj} = \hbar \sqrt{(s + k)(s - k + 1)} \delta_{j,k-1}$$
$$= \hbar \sqrt{(s - j)(s + j + 1)} \delta_{j,k-1}.$$

Therefore

$$s_+ = \hbar \begin{pmatrix} 0 & \sqrt{2s} & 0 & 0 & \cdots & 0 \\ 0 & 0 & \sqrt{2(2s-1)} & 0 & \cdots & 0 \\ 0 & 0 & 0 & \sqrt{3(2s-2)} & \cdots & 0 \\ \vdots & \vdots & \vdots & \vdots & \ddots & \vdots \\ 0 & 0 & 0 & 0 & \cdots & \sqrt{2s} \\ 0 & 0 & 0 & 0 & \cdots & 0 \end{pmatrix}.$$

Thus $s_- = (s_+)^*$.

(ii) We have

$$\mathbf{s}^2 = \frac{1}{2}(s_+ s_- + s_- s_+) + s_z^2.$$

Thus

$$(s)^2_{jk} = s(s+1)\hbar^2 \delta_{jk}.$$

Therefore s^2 is a diagonal matrix.
(iii) We find

$$e^*_{ss}s_+e_{ss} = 0$$
$$e^*_{ss}s_-e_{ss} = 0$$
$$e^*_{ss}s_ze_{ss} = \hbar s.$$

Problem 50.   The *Fibonacci numbers* are defined by the recurrence relation (linear difference equation of second order with constant coefficients)

$$s_{n+2} = s_{n+1} + s_n$$

where $n = 0, 1, \ldots$ and $s_0 = 0$, $s_1 = 1$. Write this recurrence relation in matrix form. Find s_6, s_5, and s_4.

Solution 50.   We have $s_2 = 1$. We can write

$$\begin{pmatrix} s_{n+1} & s_n \\ s_n & s_{n-1} \end{pmatrix} = \begin{pmatrix} 1 & 1 \\ 1 & 0 \end{pmatrix}^n, \qquad n = 1, 2, \ldots \quad .$$

Thus

$$\begin{pmatrix} 1 & 1 \\ 1 & 0 \end{pmatrix}^5 = \begin{pmatrix} 8 & 5 \\ 5 & 3 \end{pmatrix}.$$

It follows that $s_6 = 8$, $s_5 = 5$ and $s_4 = 3$.

Problem 51.   (i) Find four unit (column) vectors x_1, x_2, x_3, x_4 in $\mathbf{R}^3$ such that

$$x^T_j x_k = \frac{4}{3}\delta_{jk} - \frac{1}{3} = \begin{cases} 1 & \text{for } j = k \\ -1/3 & \text{for } j \neq k. \end{cases}$$

Give a geometric interpretation.
(ii) Calculate

$$\sum_{j=1}^{4} x_j.$$

(iii) Calculate

$$\sum_{j=1}^{4} x_j x^T_j.$$

Solution 51.   (i) The four vectors consist of the vectors pointing from the center of a cube to nonadjacent corners. Alternatively, one may picture these vectors as the normal vectors for the faces of the tetrahedron that is defined by the other four corners of the cube. Thus an example is

$$\mathbf{x}_1 = \frac{1}{\sqrt{3}} \begin{pmatrix} 1 \\ 1 \\ 1 \end{pmatrix}, \qquad \mathbf{x}_2 = \frac{1}{\sqrt{3}} \begin{pmatrix} 1 \\ -1 \\ -1 \end{pmatrix},$$

$$\mathbf{x}_3 = \frac{1}{\sqrt{3}} \begin{pmatrix} -1 \\ 1 \\ -1 \end{pmatrix}, \qquad \mathbf{x}_4 = \frac{1}{\sqrt{3}} \begin{pmatrix} -1 \\ -1 \\ 1 \end{pmatrix}.$$

(ii) We obviously find

$$\sum_{j=1}^{4} \mathbf{x}_j = \mathbf{0}$$

which states that the four vectors are linearly dependent.
(iii) We obtain

$$\sum_{j=1}^{4} \mathbf{x}_j \mathbf{x}_j^T = \frac{4}{3} I_3$$

where I_3 is the 3×3 identity matrix.

Problem 52. Assume that

$$A = A_1 + iA_2$$

is a nonsingular $n \times n$ matrix, where A_1 and A_2 are real $n \times n$ matrices. Assume that A_1 is also nonsingular. Find the inverse of A using the inverse of A_1.

Solution 52. We have the identity

$$(A_1 + iA_2)(I_n - iA_1^{-1}A_2) \equiv A_1 + A_2 A_1^{-1} A_2 .$$

Thus we find the inverse

$$A^{-1} = (A_1 + A_2 A_1^{-1} A_2)^{-1} - iA_1^{-1}A_2(A_1 + A_2 A_1^{-1} A_2)^{-1} .$$

Problem 53. The 8×8 matrix

$$H = \begin{pmatrix} 1 & 1 & 1 & 0 & 1 & 0 & 0 & 0 \\ 1 & 1 & 1 & 0 & -1 & 0 & 0 & 0 \\ 1 & 1 & -1 & 0 & 0 & 1 & 0 & 0 \\ 1 & 1 & -1 & 0 & 0 & -1 & 0 & 0 \\ 1 & -1 & 0 & 1 & 0 & 0 & 1 & 0 \\ 1 & -1 & 0 & 1 & 0 & 0 & -1 & 0 \\ 1 & -1 & 0 & -1 & 0 & 0 & 0 & 1 \\ 1 & -1 & 0 & -1 & 0 & 0 & 0 & -1 \end{pmatrix}$$

plays a role in the *discrete wavelet transform*.
(i) Show that the matrix is invertible without calculating the determinant.
(ii) Find the inverse.

Solution 53. (i) All the column vectors in the matrix H are nonzero. All the pairwise scalar products of the column vectors are 0. Thus the matrix has maximum rank, i.e. $\text{rank} H = 8$ and the column vectors form a basis (not normalized) in $\mathbf{R}^n$.
(ii) The inverse matrix is given by

$$H^{-1} = \frac{1}{8} \begin{pmatrix} 1 & 1 & 1 & 1 & 1 & 1 & 1 & 1 \\ 1 & 1 & 1 & 1 & -1 & -1 & -1 & -1 \\ 2 & 2 & -2 & -2 & 0 & 0 & 0 & 0 \\ 0 & 0 & 0 & 0 & 2 & 2 & -2 & -2 \\ 4 & -4 & 0 & 0 & 0 & 0 & 0 & 0 \\ 0 & 0 & 4 & -4 & 0 & 0 & 0 & 0 \\ 0 & 0 & 0 & 0 & 4 & -4 & 0 & 0 \\ 0 & 0 & 0 & 0 & 0 & 0 & 4 & -4 \end{pmatrix}.$$

Problem 54. Let A and B be $n \times n$ matrices over $\mathbf{R}$. Assume that $A \neq B$, $A^3 = B^3$ and $A^2 B = B^2 A$. Is $A^2 + B^2$ invertible?

Solution 54. We have

$$(A^2 + B^2)(A - B) = A^3 - B^3 - A^2 B + B^2 A = 0_n.$$

Since $A \neq B$, we can conclude that $A^2 + B^2$ is not invertible.

Problem 55. Let A be a positive definite $n \times n$ matrix over $\mathbf{R}$. Let $\mathbf{x} \in \mathbf{R}$. Show that $A + \mathbf{x}\mathbf{x}^T$ is also positive definite.

Solution 55. For all vector $\mathbf{y} \in \mathbf{R}^n$, $\mathbf{y} \neq \mathbf{0}$, we have $\mathbf{y}^T A \mathbf{y} > 0$. We have

$$\mathbf{y}^T (\mathbf{x}\mathbf{x}^T)\mathbf{y} = (\mathbf{y}^T\mathbf{x})(\mathbf{x}^T\mathbf{y}) = \left(\sum_{j=1}^n y_j x_j\right)\left(\sum_{j=1}^n x_j y_j\right) = \left(\sum_{j=1}^n x_j y_j\right)^2 \geq 0$$

and therefore we have $\mathbf{y}^T (A + \mathbf{x}\mathbf{x}^T)\mathbf{y} > 0$ for all $\mathbf{y} \in \mathbf{R}^n$, $\mathbf{y} \neq \mathbf{0}$.

Problem 56. Let A, B be $n \times n$ matrices over $\mathbf{C}$. The matrix A is called *similar* to the matrix B if there is an $n \times n$ invertible matrix S such that

$$A = S^{-1} B S.$$

If A is similar to B, then B is also similar to A, since $B = SAS^{-1}$.
(i) Consider the two matrices

$$A = \begin{pmatrix} 1 & 0 \\ 2 & 1 \end{pmatrix}, \qquad B = \begin{pmatrix} 1 & 0 \\ 0 & 1 \end{pmatrix}.$$

Are the matrices similar?
(ii) Consider the two matrices

$$C = \begin{pmatrix} 1 & 0 \\ 0 & -1 \end{pmatrix}, \qquad D = \begin{pmatrix} 0 & 1 \\ 1 & 0 \end{pmatrix}.$$

Are the matrices similar?

Solution 56. (i) From $A = S^{-1}BS$ we obtain $SA = BS$. Let

$$S = \begin{pmatrix} s_{11} & s_{12} \\ s_{21} & s_{22} \end{pmatrix}.$$

Then from $SA = BS$ we obtain

$$\begin{pmatrix} s_{11} + 2s_{12} & s_{12} \\ s_{21} + 2s_{22} & s_{22} \end{pmatrix} = \begin{pmatrix} s_{11} & s_{12} \\ s_{21} & s_{22} \end{pmatrix}.$$

It follows that $s_{12} = 0$ and $s_{22} = 0$. Thus S is not invertible and therefore A and B are not similar.
(ii) The matrices C and D are similar. We find

$$s_{11} = s_{21}, \qquad s_{12} = -s_{22}.$$

Since S must be invertible, all four matrix elements are nonzero. For example, we can select

$$S = \frac{1}{\sqrt{2}} \begin{pmatrix} 1 & 1 \\ 1 & -1 \end{pmatrix}.$$

Chapter 2

Linear Equations

Let A be an $m \times n$ matrix over a field $\mathcal{F}$. Let $b_1, \ldots, b_m$ be elements of the field $\mathcal{F}$. The system of equations

$$a_{11}x_1 + a_{12}x_2 + \cdots + a_{1n}x_n = b_1$$
$$a_{21}x_1 + a_{22}x_2 + \cdots + a_{2n}x_n = b_2$$
$$\vdots \quad \vdots$$
$$a_{m1}x_1 + a_{m2}x_2 + \cdots + a_{mn}x_n = b_m$$

is called a system of linear equations. We also write $A\mathbf{x} = \mathbf{b}$, where $\mathbf{x}$ and $\mathbf{b}$ are considered as column vectors. The system is said to be homogeneous if all the numbers $b_1, \ldots, b_m$ are equal to 0. The number n is called the number of unknowns, and m is called the number of equations. The system of homogeneous equations also admits the trivial solution $x_1 = x_2 = \cdots = x_n = 0$.

A system of homogeneous equations of m linear equations in n unknowns with $n > m$ admits a non-trivial solution. An underdetermined linear system is either inconsistent or has infinitely many solutions.

An important special case is $m = n$. Then for the system of linear equations $A\mathbf{x} = \mathbf{b}$ we investigate the cases A^{-1} exists and A^{-1} does not exist. If A^{-1} exists we can write the solution as $\mathbf{x} = A^{-1}\mathbf{b}$.

If $m > n$, then we have an overdetermined system and it can happen that no solution exists. One solves these problems in the least-square sense.

34

Problem 1. Let

$$A = \begin{pmatrix} 1 & 1 \\ 2 & -1 \end{pmatrix}, \qquad b = \begin{pmatrix} 1 \\ 5 \end{pmatrix}.$$

Find the solutions of the system of linear equations

$$Ax = b.$$

Solution 1. Since A is invertible we have the unique solution

$$x = A^{-1}b.$$

From the equations

$$x_1 + x_2 = 1$$
$$2x_1 - x_2 = 5$$

we obtain by addition of the two equations $3x_1 = 6$ and thus $x_1 = 2$. It follows that $x_2 = -1$.

Problem 2. Let

$$A = \begin{pmatrix} 1 & 1 \\ 2 & 2 \end{pmatrix}, \qquad b = \begin{pmatrix} 3 \\ \alpha \end{pmatrix}$$

where $\alpha \in \mathbf{R}$. What is the condition on α so that there is a solution of the equation $Ax = b$?

Solution 2. From $Ax = b$ we obtain

$$x_1 + x_2 = 3$$
$$2x_1 + 2x_2 = \alpha.$$

Multiplying the first equation by 2 and then subtracting from the second equation yields $6 = \alpha$. Thus if $\alpha \neq 6$ there is no solution. If $\alpha = 6$ the line $x_1 + x_2 = 3$ is the solution.

Problem 3. (i) Find all solutions of the system of linear equations

$$\begin{pmatrix} \cos\theta & -\sin\theta \\ -\sin\theta & -\cos\theta \end{pmatrix} \begin{pmatrix} x_1 \\ x_2 \end{pmatrix} = \begin{pmatrix} x_1 \\ x_2 \end{pmatrix}, \qquad \theta \in \mathbf{R}.$$

(ii) What type of equation is this?

Solution 3.   (i) We obtain

$$\begin{pmatrix} x_1 \\ x_2 \end{pmatrix} = \begin{pmatrix} \cos(\theta/2) \\ -\sin(\theta/2) \end{pmatrix}$$

where we used the identities

$$\sin\theta \equiv 2\sin(\theta/2)\cos(\theta/2), \qquad \cos\theta \equiv 2\cos^2(\theta/2) - 1.$$

(ii) This is an eigenvalue equation with eigenvalue 1.

Problem 4.   Let $A \in \mathbf{R}^{n \times n}$ and $\mathbf{x}, \mathbf{b} \in \mathbf{R}^n$. Consider the linear equation $A\mathbf{x} = \mathbf{b}$. Show that it can be written as $\mathbf{x} = T\mathbf{x}$, i.e., find $T\mathbf{x}$.

Solution 4.   Let $C = I_n - A$. Then we can write

$$\mathbf{x} = C\mathbf{x} + \mathbf{b}.$$

Thus $\mathbf{x} = T\mathbf{x}$ with $T\mathbf{x} := C\mathbf{x} + \mathbf{b}$.

Problem 5.   If the system of linear equations $A\mathbf{x} = \mathbf{b}$ admits no solution we call the equations inconsistent. If there is a solution, the equations are called consistent. Let $A\mathbf{x} = \mathbf{b}$ be a system of m linear equations in n unknowns and suppose that the rank of A is m. Show that in this case $A\mathbf{x} = \mathbf{b}$ is consistent.

Solution 5.   Since $[A|\mathbf{b}]$ is an $m \times (n+1)$ matrix we have $m \geq \text{rank}[A|\mathbf{b}]$. We have $\text{rank}[A|\mathbf{b}] \geq \text{rank}A$ and by assumption $\text{rank}A = m$. Thus

$$m \geq \text{rank}[A|\mathbf{b}] \geq \text{rank}A = m.$$

Hence $\text{rank}[A|\mathbf{b}] = m$ and therefore the system of equations $A\mathbf{x} = \mathbf{b}$ are consistent.

Problem 6.   Show that the *curve fitting problem*

j	0	1	2	3	4
t_j	−1.0	−0.5	0.0	0.5	1.0
y_j	1.0	0.5	0.0	0.5	2.0

by a quadratic polynomial of the form

$$p(t) = a_2 t^2 + a_1 t + a_0$$

leads to an overdetermined linear system.

Solution 6. From the interpolation conditions $p(t_j) = y_j$ with $j = 0, 1, \ldots, 4$ we obtain the overdetermined linear system

$$\begin{pmatrix} 1 & t_0 & t_0^2 \\ 1 & t_1 & t_1^2 \\ 1 & t_2 & t_2^2 \\ 1 & t_3 & t_3^2 \\ 1 & t_4 & t_4^2 \end{pmatrix} \begin{pmatrix} a_0 \\ a_1 \\ a_2 \end{pmatrix} = \begin{pmatrix} y_0 \\ y_1 \\ y_2 \\ y_3 \\ y_4 \end{pmatrix}.$$

Problem 7. Consider the overdetermined linear system $Ax = b$. Find an $\hat{x}$ such that

$$\|A\hat{x} - b\|_2 = \min_x \|Ax - b\|_2 \equiv \min_x \|r(x)\|_2$$

with the *residual vector* $r(x) := b - Ax$ and $\|.\|_2$ denotes the Euclidean norm.

Solution 7. From

$$\|r(x)\|_2^2 = r^T r = (b - Ax)^T (b - Ax) = b^T b - 2x^T A^T b + x^T A^T Ax$$

where we used that $x^T A^T b = b^T Ax$, and the necessary condition

$$\nabla \|r(x)\|_2^2|_{x=\hat{x}} = 0$$

we obtain

$$A^T A\hat{x} - A^T b = 0.$$

This system is called *normal equations*. We can also write this system as

$$A^T (b - A\hat{x}) \equiv A^T r(\hat{x}) = 0.$$

This justifies the name normal equations.

Problem 8. Consider the overdetermined linear system $Ax = b$ with

$$A = \begin{pmatrix} 1 & 1 \\ 1 & 2 \\ 1 & 3 \\ 1 & 4 \\ 1 & 5 \\ 1 & 6 \\ 1 & 7 \\ 1 & 8 \\ 1 & 9 \\ 1 & 10 \end{pmatrix}, \quad x = \begin{pmatrix} x_1 \\ x_2 \end{pmatrix}, \quad b = \begin{pmatrix} 444 \\ 458 \\ 478 \\ 493 \\ 506 \\ 516 \\ 523 \\ 531 \\ 543 \\ 571 \end{pmatrix}.$$

Solve this linear system in the least squares sense (see previous problem) by the normal equations method.

Solution 8.   From the normal equations

$$A^T A\hat{x} = A^T b$$

we obtain

$$\begin{pmatrix} 10 & 55 \\ 55 & 385 \end{pmatrix} \begin{pmatrix} \hat{x}_1 \\ \hat{x}_2 \end{pmatrix} = \begin{pmatrix} 5063 \\ 28898 \end{pmatrix}$$

with $\det(A^T A) \neq 0$. The solution is approximately

$$\begin{pmatrix} \hat{x}_1 \\ \hat{x}_2 \end{pmatrix} = \begin{pmatrix} 436.2 \\ 12.7455 \end{pmatrix} .$$

Problem 9.   An underdetermined linear system is either inconsistent or has infinitely many solutions. Consider the underdetermined linear system

$$Hx = y$$

where H is an $n \times m$ matrix with $m > n$ and

$$x = \begin{pmatrix} x_1 \\ x_2 \\ \vdots \\ x_m \end{pmatrix}, \qquad y = \begin{pmatrix} y_1 \\ y_2 \\ \vdots \\ y_n \end{pmatrix} .$$

We assume that $Hx = y$ has infinitely many solutions. Let P be the $n \times m$ matrix

$$P = \begin{pmatrix} 1 & 0 & \cdots & 0 & 0 & \cdots & 0 \\ 0 & 1 & \cdots & 0 & 0 & \cdots & 0 \\ \vdots & \vdots & \ddots & \vdots & \vdots & & \vdots \\ 0 & 0 & \cdots & 1 & 0 & \cdots & 0 \end{pmatrix} .$$

We define $\hat{x} := Px$. Find

$$\min_{x} \| Px - y \|_2^2$$

subject to the constraint $\| Hx - y \|_2^2 = 0$. We assume that $(\lambda H^T H + P^T P)^{-1}$ exists for all $\lambda > 0$. Apply the *Lagrange multiplier method*.

Solution 9.   We have

$$V(x) = \| Px - y \|_2^2 + \lambda \| Hx - y \|_2^2$$

where λ is the Lagrange multiplier. Thus $V(\mathbf{x}) \to$ min if λ is sufficiently large. The derivative of $V(\mathbf{x})$ with respect to the unknown $\mathbf{x}$ is

$$\frac{\partial}{\partial \mathbf{x}} V(\mathbf{x}) = 2\lambda H^T (H\mathbf{x} - \mathbf{y}) + 2P^T (P\mathbf{x} - \mathbf{y}).$$

Thus

$$(\lambda H^T H + P^T P)\mathbf{x} = (\lambda H^T + P^T)\mathbf{y}.$$

It follows that

$$\hat{\mathbf{x}} = (\lambda H^T H + P^T P)^{-1} (\lambda H + P)^T \mathbf{y}.$$

Problem 10. Show that solving the system of nonlinear equations with the unknowns x_1, x_2, x_3, x_4

$$(x_1 - 1)^2 + (x_2 - 2)^2 + x_3^2 = a^2 (x_4 - b_1)^2$$
$$(x_1 - 2)^2 + x_2^2 + (x_3 - 2)^2 = a^2 (x_4 - b_2)^2$$
$$(x_1 - 1)^2 + (x_2 - 1)^2 + (x_3 - 1)^2 = a^2 (x_4 - b_3)^2$$
$$(x_1 - 2)^2 + (x_2 - 1)^2 + x_3^2 = a^2 (x_4 - b_4)^2$$

leads to a linear underdetermined system. Solve this system with respect to x_1, x_2 and x_3.

Solution 10. Expanding all the squares and rearranging that the linear terms are on the left-hand side, yields

$$2x_1 + 4x_2 - 2a^2 b_1 x_4 = 5 - a^2 b_1^2 + x_1^2 + x_2^2 + x_3^2 - a^2 x_4^2$$
$$4x_1 + 4x_3 - 2a^2 b_2 x_4 = 8 - a^2 b_2^2 + x_1^2 + x_2^2 + x_3^2 - a^2 x_4^2$$
$$2x_1 + 2x_2 + 2x_3 - 2a^2 b_3 x_4 = 3 - a^2 b_3^2 + x_1^2 + x_2^2 + x_3^2 - a^2 x_4^2$$
$$4x_1 + 2x_2 - 2a^2 b_4 x_4 = 5 - a^2 b_4^2 + x_1^2 + x_2^2 + x_3^2 - a^2 x_4^2.$$

The quadratic terms in all the equations are the same. Thus by subtracting the first equation from each of the other three, we obtain an underdetermined system of three linear equations

$$2x_1 - 4x_2 + 4x_3 - 2a^2 (b_2 - b_1) x_4 = 3 + a^2 (b_1^2 - b_2^2)$$
$$-2x_2 + 2x_3 - 2a^2 (b_3 - b_1) x_4 = -2 + a^2 (b_1^2 - b_3^2)$$
$$2x_1 - 2x_2 - 2a^2 (b_4 - b_1) x_4 = a^2 (b_1^2 - b_4^2).$$

Solving these equations we obtain

$$x_1 = a^2(b_1 + b_2 - 2b_3)x_4 + \frac{a^2}{2}(-b_1^2 - b_2^2 + 2b_3^2) + \frac{7}{2}$$

$$x_2 = a^2(2b_1 + b_2 - 2b_3 - b_4)x_4 + \frac{a^2}{2}(-2b_1^2 - b_2^2 + b_3^2 + b_4^2) + \frac{7}{2}$$

$$x_3 = a^2(b_1 + b_2 - b_3 - b_4)x_4 + \frac{a^2}{2}(-b_1^2 - b_2^2 + b_3^2 + b_4^2) + \frac{5}{2}.$$

Inserting these solutions into one of the nonlinear equations provides a quadratic equation for x_4. Such a system of equations plays a role in the *Global Positioning System* (GPS), where x_4 plays the role of time.

Problem 11.   *Kirchhoff's current law* states that the algebraic sum of all the currents flowing into a junction is 0. *Kirchhoff's voltage law* states that the algebraic sum of all the voltages around a closed circuit is 0. Use Kirchhoff's laws and *Ohm's law* $(V = RI)$ to setting up the system of linear equations for the circuit depicted in the figure.

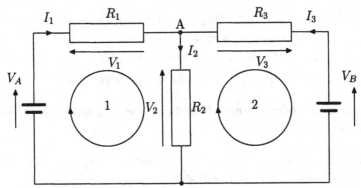

Given the voltages V_A, V_B and the resistors R_1, R_2, R_3. Find I_1, I_2, I_3.

Solution 11.   From Kirchhoff's voltage law we find for loop 1 and loop 2 that

$$V_1 + V_2 = V_A$$
$$V_2 + V_3 = V_B.$$

From Kirchhoff's current law we obtain for node A

$$I_1 - I_2 + I_3 = 0.$$

Thus the linear system can be written in matrix form

$$\begin{pmatrix} R_1 & R_2 & 0 \\ 0 & R_2 & R_3 \\ 1 & -1 & 1 \end{pmatrix} \begin{pmatrix} I_1 \\ I_2 \\ I_3 \end{pmatrix} = \begin{pmatrix} V_A \\ V_B \\ 0 \end{pmatrix}.$$

The determinant of the matrix on the left-hand side is given by

$$R_1 R_2 + R_2 R_3 + R_1 R_3 .$$

Since $R_1, R_2, R_3 > 0$ the inverse matrix exists. The solution is given by

$$I_1 = \frac{(R_2 + R_3)V_A - R_2 V_B}{R_1 R_2 + R_1 R_3 + R_2 R_3}$$

$$I_2 = \frac{R_3 V_A + R_1 V_B}{R_1 R_2 + R_2 R_3 + R_1 R_3}$$

$$I_3 = \frac{-R_2 V_A + (R_1 + R_2)V_B}{R_1 R_2 + R_2 R_3 + R_1 R_3} .$$

Problem 12. Let A be an $m \times n$ matrix over $\mathbf{R}$. We define

$$N_A := \{ \mathbf{x} \in \mathbf{R}^n \ : \ A\mathbf{x} = \mathbf{0} \} .$$

N_A is called the *kernel* of A and

$$\nu(A) := \dim(N_A)$$

is called the *nullity* of A. If N_A only contains the zero vector, then $\nu(A) = 0$.
(i) Let

$$A = \begin{pmatrix} 1 & 2 & -1 \\ 2 & -1 & 3 \end{pmatrix} .$$

Find N_A and $\nu(A)$.
(ii) Let

$$A = \begin{pmatrix} 2 & -1 & 3 \\ 4 & -2 & 6 \\ -6 & 3 & -9 \end{pmatrix} .$$

Find N_A and $\nu(A)$.

Solution 12. (i) From

$$\begin{pmatrix} 1 & 2 & -1 \\ 2 & -1 & 3 \end{pmatrix} \begin{pmatrix} x_1 \\ x_2 \\ x_3 \end{pmatrix} = \begin{pmatrix} 0 \\ 0 \end{pmatrix}$$

we find the system of linear equations

$$x_1 + 2x_2 - x_3 = 0$$
$$2x_1 - x_2 + 3x_3 = 0 .$$

Eliminating x_3 yields

$$x_1 = -x_2 .$$

It follows that $x_3 = -x_1$. Thus N_A is spanned by the vector

$$\begin{pmatrix} -1 \\ 1 \\ 1 \end{pmatrix}.$$

Therefore $\nu(A) = 1$.

(ii) From

$$\begin{pmatrix} 2 & -1 & 3 \\ 4 & -2 & 6 \\ -6 & 3 & -9 \end{pmatrix} \begin{pmatrix} x_1 \\ x_2 \\ x_3 \end{pmatrix} = \begin{pmatrix} 0 \\ 0 \\ 0 \end{pmatrix}$$

we find the system of linear equations

$$2x_1 - x_2 + 3x_3 = 0$$
$$4x_1 - 2x_2 + 6x_3 = 0$$
$$-6x_1 + 3x_2 - 9x_3 = 0.$$

The three equation are the same. Thus from $2x_1 - x_2 + 3x_3 = 0$ we find that

$$\left\{ \begin{pmatrix} 1 \\ 2 \\ 0 \end{pmatrix}, \begin{pmatrix} 0 \\ 3 \\ 1 \end{pmatrix} \right\}$$

is a basis for N_A and $\nu(A) = 2$.

Problem 13.   Let V be a vector space over a field $\mathcal{F}$. Let W be a subspace of V. We define an *equivalence relation* $\sim$ on V by stating that $v_1 \sim v_2$ if $v_1 - v_2 \in W$. The *quotient space* V/W is the set of equivalence classes $[v]$ where $v_1 - v_2 \in W$. Thus we can say that v_1 is equivalent to v_2 modulo W if $v_1 = v_2 + w$ for some $w \in W$. Let

$$V = \mathbf{R}^2 = \left\{ \begin{pmatrix} x_1 \\ x_2 \end{pmatrix} : x_1, x_2 \in \mathbf{R} \right\}$$

and the subspace

$$W = \left\{ \begin{pmatrix} x_1 \\ 0 \end{pmatrix} : x_1 \in \mathbf{R} \right\}.$$

(i) Is

$$\begin{pmatrix} 3 \\ 0 \end{pmatrix} \sim \begin{pmatrix} 1 \\ 0 \end{pmatrix}, \quad \begin{pmatrix} 4 \\ 1 \end{pmatrix} \sim \begin{pmatrix} -3 \\ 1 \end{pmatrix}, \quad \begin{pmatrix} 3 \\ 0 \end{pmatrix} \sim \begin{pmatrix} 4 \\ 1 \end{pmatrix} ?$$

(ii) Give the quotient space V/W.

Solution 13.   (i) We have

$$\begin{pmatrix} 3 \\ 0 \end{pmatrix} \sim \begin{pmatrix} 1 \\ 0 \end{pmatrix} \quad \text{since} \quad \begin{pmatrix} 3 \\ 0 \end{pmatrix} - \begin{pmatrix} 1 \\ 0 \end{pmatrix} = \begin{pmatrix} 2 \\ 0 \end{pmatrix} \in W$$

$$\begin{pmatrix} 4 \\ 1 \end{pmatrix} \sim \begin{pmatrix} -3 \\ 1 \end{pmatrix} \quad \text{since} \quad \begin{pmatrix} 4 \\ 1 \end{pmatrix} - \begin{pmatrix} -3 \\ 1 \end{pmatrix} = \begin{pmatrix} 7 \\ 0 \end{pmatrix} \in W$$

$$\begin{pmatrix} 3 \\ 0 \end{pmatrix} \not\sim \begin{pmatrix} 4 \\ 1 \end{pmatrix} \quad \text{since} \quad \begin{pmatrix} 3 \\ 0 \end{pmatrix} - \begin{pmatrix} 4 \\ 1 \end{pmatrix} = \begin{pmatrix} -1 \\ -1 \end{pmatrix} \notin W.$$

(ii) Thus the elements of the quotient space consists of straight lines parallel to the x_1 axis.

Problem 14. Let $x_1, x_2, x_3 \in \mathbf{Z}$. Find all solutions of the system of linear equations

$$7x_1 + 5x_2 - 5x_3 = 8$$
$$17x_1 + 10x_2 - 15x_3 = -42.$$

Find all positive solutions.

Solution 14. Eliminating x_2 yields $3x_1 - 5x_3 = -58$ or

$$3x_1 \equiv -58 \pmod{5}.$$

The solution is $x_1 \equiv 4 \pmod{5}$ or

$$x_1 = 4 + 5s, \qquad s \in \mathbf{Z}.$$

Thus using $3x_1 - 5x_3 = -58$ we obtain

$$x_3 = 14 + 3s$$

and using $7x_1 + 5x_2 - 5x_3 = 8$ we find

$$x_2 = 10 - 4s.$$

For x_2 positive we have $s \leq 2$, for x_1 positive we have $0 \leq s$ and x_3 remains positive. Thus the solution set for positive x_1, x_2, x_3 is

$$(4, 10, 14), \quad (9, 6, 17), \quad (14, 2, 20).$$

We can write

$$\begin{pmatrix} x_1 \\ x_2 \\ x_3 \end{pmatrix} = \begin{pmatrix} 4 \\ 10 \\ 14 \end{pmatrix} + s \begin{pmatrix} 5 \\ -4 \\ 3 \end{pmatrix}$$

where $s = 0, 1, 2$.

Problem 15. Consider the inhomogeneous linear integral equation

$$\int_0^1 (\alpha_1(x)\beta_1(y) + \alpha_2(x)\beta_2(y))\varphi(y)dy + f(x) = \varphi(x) \tag{1}$$

for the unknown function φ, $f(x) = x$ and

$$\alpha_1(x) = x, \quad \alpha_2(x) = \sqrt{x}, \quad \beta_1(y) = y, \quad \beta_2(y) = \sqrt{y}.$$

Thus α_1 and α_2 are continuous in $[0, 1]$ and likewise for β_1 and β_2. We define

$$B_1 := \int_0^1 \beta_1(y)\varphi(y)dy, \qquad B_2 := \int_0^1 \beta_2(y)\varphi(y)dy$$

and

$$a_{\mu\nu} := \int_0^1 \beta_\mu(y)\alpha_\nu(y)dy, \qquad b_\mu := \int_0^1 \beta_\mu(y)f(y)dy$$

where $\mu, \nu = 1, 2$. Show that the integral equation can be cast into a system of linear equations for B_1 and B_2. Solve this system of linear equations and thus find a solution of the integral equation.

Solution 15.   Using B_1 and B_2 equation (1) can be written as

$$\varphi(x) = \alpha_1(x)B_1 + \alpha_2(x)B_2 + f(x) \tag{2}$$

or

$$\varphi(y) = \alpha_1(y)B_1 + \alpha_2(y)B_2 + f(y). \tag{3}$$

We insert (2) into the left-hand side and (3) into the right-hand side of

$$\varphi(x) = \alpha_1(x)\int_0^1 \beta_1(y)\varphi(y)dy + \alpha_2(x)\int_0^1 \beta_2(y)\varphi(y)dy + f(x).$$

Using $a_{\mu\nu}$ and b_μ we find

$$\alpha_1(x)B_1 + \alpha_2(x)B_2 = \alpha_1(x)B_1a_{11} + \alpha_1(x)B_2a_{12} + \alpha_1(x)b_1$$
$$+\alpha_2(x)B_1a_{21} + \alpha_2(x)B_2a_{22} + \alpha_2(x)b_2.$$

Now α_1 and α_2 are linearly independent. Comparing the coefficients of α_1 and α_2 we obtain the system of linear equations for B_1 and B_2

$$B_1 = a_{11}B_1 + a_{12}B_2 + b_1$$
$$B_2 = a_{21}B_1 + a_{22}B_2 + b_2$$

or in matrix form

$$\begin{pmatrix} 1 - a_{11} & -a_{12} \\ -a_{21} & 1 - a_{22} \end{pmatrix} \begin{pmatrix} B_1 \\ B_2 \end{pmatrix} = \begin{pmatrix} b_1 \\ b_2 \end{pmatrix}.$$

Since

$$a_{11} = \int_0^1 y^2 dy = \frac{1}{3}, \qquad a_{12} = \int_0^1 y^{3/2} dy = \frac{2}{5}$$

$$a_{21} = \int_0^1 y^{3/2}\,dy = \frac{2}{5}, \qquad a_{22} = \int_0^1 y\,dy = \frac{1}{2}.$$

$$b_1 = \int_0^1 y^2\,dy = \frac{1}{3}, \qquad b_2 = \int_0^1 y^{2/3}\,dy = \frac{2}{5}.$$

Therefore

$$\begin{pmatrix} 2/3 & -2/5 \\ -2/5 & 1/2 \end{pmatrix} \begin{pmatrix} B_1 \\ B_2 \end{pmatrix} = \begin{pmatrix} 1/3 \\ 2/5 \end{pmatrix}$$

with the unique solution

$$B_1 = \frac{49}{26}, \qquad B_2 = \frac{30}{13}.$$

Since

$$\varphi(x) = \alpha_1(x)B_1 + \alpha_2(x)B_2 + x$$

we obtain the solution of the integral equation

$$\varphi(x) = \frac{75}{26}x + \frac{30}{13}\sqrt{x}.$$

Chapter 3

Determinants and Traces

A function on $n \times n$ matrices $\det : \mathbf{C}^{n \times n} \to \mathbf{C}$ is called a determinant function if and only if it satisfies the following conditions:

1) det is linear in each row if the other rows of the matrix are held fixed.
2) If the $n \times n$ matrix A has two identical rows then $\det A = 0$.
3) If I_n is the $n \times n$ identity matrix, then $\det I_n = 1$.

Let A, B be $n \times n$ matrices and $c \in \mathbf{C}$. Then we have

$$\det(AB) = \det A \det B, \qquad \det(cA) = c^n \det A.$$

The determinant of A is the product of the eigenvalues of A

$$\det A = \lambda_1 \cdot \lambda_2 \cdot \ldots \cdot \lambda_n.$$

Let A be an $n \times n$ matrix. Then the *trace* is defined as

$$\mathrm{tr} A := \sum_{j=1}^{n} a_{jj}.$$

The trace is independent of the underlying basis. The trace is the sum of the eigenvalues of A, i.e.

$$\mathrm{tr} A = \sum_{j=1}^{n} \lambda_j.$$

The trace and determinant of a square matrix A are related by the identity

$$\det \exp(A) \equiv \exp(\mathrm{tr} A).$$

46

Problem 1. Let A be a 2×2 matrix over $\mathbf{C}$. Using the *Cayley-Hamilton theorem* show that

$$(\operatorname{tr} A)^2 = \operatorname{tr}(A^2) + 2 \det(A). \tag{1}$$

Cayley-Hamilton theorem. Let A be an $n \times n$ matrix over a field (in our case $\mathbf{C}$) with characteristic polynomial $p_A(\lambda) = \det(\lambda I_n - A)$, where I_n is the $n \times n$ identity matrix. Then $p_A(A)$ is the $n \times n$ zero matrix.

Solution 1. Thus for $n = 2$ we have

$$A^2 - A \operatorname{tr} A + I_2 \det(A) = 0.$$

Taking the trace of the left and right hand side and $\operatorname{tr} I_2 = 2$ we obtain (1).

Problem 2. Consider the 2×2 matrix

$$A = \begin{pmatrix} 0 & 1 \\ 0 & 0 \end{pmatrix}.$$

Can we find an invertible 2×2 matrix Q such that $Q^{-1}AQ$ is a diagonal matrix?

Solution 2. The answer is no. Let

$$\tilde{A} = Q^{-1}AQ \tag{1}$$

with

$$\tilde{A} = \begin{pmatrix} a & 0 \\ 0 & b \end{pmatrix}$$

where $a, b \in \mathbf{C}$. Taking the trace of (1) we find

$$a + b = \operatorname{tr}(Q^{-1}AQ) = \operatorname{tr} A = 0.$$

Taking the determinant of (1) we obtain

$$ab = \det(Q^{-1}AQ) = \det(Q^{-1}) \det(A) \det(Q) = \det A = 0.$$

Thus from $a + b = 0$ and $ab = 0$ we find $a = b = 0$. However from (1) it also follows that

$$Q\tilde{A}Q^{-1} = A.$$

Therefore A is the zero matrix which contradicts the assumption for the matrix A. Thus no invertible Q can be found. This means A is not diagonalizable.

Problem 3. Let A be a 2×2 matrix over $\mathbf{R}$. Assume that $\operatorname{tr} A = 0$ and $\operatorname{tr} A^2 = 0$. Can we conclude that A is the 2×2 zero matrix?

Solution 3.   No we cannot conclude that A is the 2×2 zero matrix. For example,

$$A = \begin{pmatrix} 0 & 0 \\ 1 & 0 \end{pmatrix}$$

satisfies $\operatorname{tr} A = 0$ and $\operatorname{tr} A^2 = 0$. What happens if we also assume that A is symmetric over $\mathbf{R}$?

Problem 4.   Consider the $(n - 1) \times (n - 1)$ matrix

$$A = \begin{pmatrix} 3 & 1 & 1 & 1 & \cdots & 1 \\ 1 & 4 & 1 & 1 & \cdots & 1 \\ 1 & 1 & 5 & 1 & \cdots & 1 \\ 1 & 1 & 1 & 6 & \cdots & 1 \\ \vdots & \vdots & \vdots & \vdots & \ddots & \vdots \\ 1 & 1 & 1 & 1 & \cdots & n+1 \end{pmatrix}.$$

Let D_n be the determinant of this matrix. Is the sequence $\{ D_n/n! \}$ bounded?

Solution 4.   If we expand the last row we obtain the recursion

$$D_n = n D_{n-1} + (n-1)!, \qquad n = 3, 4, \ldots$$

with the initial condition $D_2 = 3$. We divide by $n!$ to obtain

$$\frac{D_n}{n!} = \frac{D_{n-1}}{(n-1)!} + \frac{1}{n}.$$

Thus

$$D_n = n! \left(1 + \frac{1}{2} + \cdots + \frac{1}{n} \right).$$

Therefore the sequence $\{ D_n/n! \}$ is the nth partial sum of the *harmonic series*, which is unbounded as $n \to \infty$.

Problem 5.   For an integer $n \geq 3$, let $\theta := 2\pi/n$. Find the determinant of the $n \times n$ matrix $A + I_n$, where I_n is the $n \times n$ identity matrix and the matrix $A = (a_{jk})$ has the entries $a_{jk} = \cos(j\theta + k\theta)$ for all $j, k = 1, 2, \ldots, n$.

Solution 5.   The determinant of a square matrix is the product of its eigenvalues. We compute the determinant by calculating the eigenvalues of $I_n + A$. The eigenvalues of $I_n + A$ are obtained by adding 1 to each of the eigenvalues of A. Thus we only have to calculate the eigenvalues of A and then add 1. We show that the eigenvalues of A are $n/2, -n/2, 0, \ldots, 0$, where 0 occurs with multiplicity $n - 2$. We define column vectors $\mathbf{v}^{(m)}$,

$0 \le m \le n-1$, componentwise by $v_k^{(m)} = e^{ikm\theta}$, where $\theta = 2\pi/n$. We form a matrix from the column vectors $v^{(m)}$. Its determinant is a Vandermonde product and hence is nonzero. Thus the vectors $v^{(m)}$ form a basis in $\mathbf{C}^n$. Since $\cos z \equiv (e^{iz} + e^{-iz})/2$ for any $z \in \mathbf{C}$ we obtain

$$(A v^{(m)})_j = \sum_{k=1}^{n} \cos(j\theta + k\theta) e^{ikm\theta}$$

$$= \frac{e^{ij\theta}}{2} \sum_{k=1}^{n} e^{ik(m+1)\theta} + \frac{e^{-ij\theta}}{2} \sum_{k=1}^{n} e^{ik(m-1)\theta}$$

where $j = 1, 2, \ldots, n$. Since

$$\sum_{k=1}^{n} e^{ik\ell\theta} = 0$$

for integer ℓ unless $n|\ell$, we conclude that $A v^{(m)} = \mathbf{0}$ for $m = 0$ and for $2 \le m \le n-1$. In addition, we find that

$$(A v^{(1)})_j = \frac{n}{2} e^{-ij\theta} = \frac{n}{2} (v^{(n-1)})_j, \qquad (A v^{(n-1)})_j = \frac{n}{2} e^{ij\theta} = \frac{n}{2} (v^{(1)})_j .$$

Thus

$$A(v^{(1)} \pm v^{(n-1)}) = \pm \frac{n}{2} (v^{(1)} \pm v^{(n-1)}) .$$

Consequently

$$\{ v^{(0)}, v^{(2)}, v^{(3)}, \ldots, v^{(n-2)}, v^{(1)} + v^{(n-1)}, v^{(1)} - v^{(n-1)} \}$$

is a basis for $\mathbf{C}^n$ of eigenvectors of A with the claimed eigenvalues. Since the determinant of $I_n + A$ is the product of $(1 + \lambda)$ over all eigenvalues λ of A, we obtain

$$\det(I_n + A) = (1 + n/2)(1 - n/2) = 1 - n^2/4 .$$

Problem 6. Let α, β, γ, δ be real numbers.
(i) Is the matrix

$$U = e^{i\alpha} \begin{pmatrix} e^{-i\beta/2} & 0 \\ 0 & e^{i\beta/2} \end{pmatrix} \begin{pmatrix} \cos(\gamma/2) & -\sin(\gamma/2) \\ \sin(\gamma/2) & \cos(\gamma/2) \end{pmatrix} \begin{pmatrix} e^{-i\delta/2} & 0 \\ 0 & e^{i\delta/2} \end{pmatrix}$$

unitary?
(ii) What the determinant of U?

Solution 6. (i) Each of the three matrices on the right-hand side is unitary and $e^{i\alpha}$ is unitary. The product of two unitary matrices is again a unitary matrix. Thus U is unitary.

(ii) The determinat of each of the three matrices on the right-hand side is 1. Thus $\det(U) = e^{2i\alpha}$.

Problem 7. Let A and B be two $n \times n$ matrices over $\mathbf{C}$. If there exists a non-singular $n \times n$ matrix X such that

$$A = XBX^{-1}$$

then A and B are said to be *similar matrices*. Show that the spectra (eigenvalues) of two similar matrices are equal.

Solution 7. We have

$$\begin{aligned}
\det(A - \lambda I_n) &= \det(XBX^{-1} - X\lambda I_n X^{-1}) \\
&= \det(X(B - \lambda I_n)X^{-1}) \\
&= \det(X)\det(B - \lambda I_n)\det(X^{-1}) \\
&= \det(B - \lambda I_n).
\end{aligned}$$

Problem 8. Let A be an $n \times n$ matrix. Assume that the inverse matrix of A exists. The inverse matrix can be calculated as follows (*Csanky's algorithm*). Let

$$p(x) := \det(xI_n - A) \tag{1}$$

where I_n is the $n \times n$ unit matrix. The roots are, by definition, the eigenvalues $\lambda_1, \lambda_2, \cdots, \lambda_n$ of A. We write

$$p(x) = x^n + c_1 x^{n-1} + \cdots + c_{n-1}x + c_n \tag{2}$$

where

$$c_n = (-1)^n \det A.$$

Since A is nonsingular we have $c_n \neq 0$ and vice versa. The *Cayley-Hamilton theorem* states that

$$p(A) = A^n + c_1 A^{n-1} + \cdots + c_{n-1}A + c_n I_n = 0_n. \tag{3}$$

Multiplying this equation with A^{-1} we obtain

$$A^{-1} = \frac{1}{-c_n}(A^{n-1} + c_1 A^{n-2} + \cdots + c_{n-1}I_n). \tag{4}$$

If we have the coefficients c_j we can calculate the inverse matrix A. Let

$$s_k := \sum_{j=1}^{n} \lambda_j^k.$$

Then the s_j and c_j satisfy the following $n \times n$ lower triangular system of linear equations

$$
\begin{pmatrix}
1 & 0 & 0 & \cdots & 0 \\
s_1 & 2 & 0 & \cdots & 0 \\
s_2 & s_1 & 3 & \cdots & 0 \\
\vdots & \vdots & \vdots & \ddots & \vdots \\
s_{n-1} & s_{n-2} & \cdots & s_1 & n
\end{pmatrix}
\begin{pmatrix}
c_1 \\ c_2 \\ c_3 \\ \vdots \\ c_n
\end{pmatrix}
=
\begin{pmatrix}
-s_1 \\ -s_2 \\ -s_3 \\ \vdots \\ -s_n
\end{pmatrix} .
$$

Since

$$
\mathrm{tr}(A^k) = \lambda_1^k + \lambda_2^k + \cdots + \lambda_n^k = s_k
$$

we find s_k for $k = 1, 2, \ldots, n$. Thus we can solve the linear equation for c_j. Finally, using (4) we obtain the inverse matrix of A. Apply Csanky's algorithm to the 3×3 matrix

$$
A = \begin{pmatrix}
1 & 0 & 1 \\
1 & 0 & 0 \\
0 & 1 & 1
\end{pmatrix} .
$$

Solution 8. Since

$$
A^2 = \begin{pmatrix}
1 & 1 & 2 \\
1 & 0 & 1 \\
1 & 1 & 1
\end{pmatrix} , \qquad
A^3 = \begin{pmatrix}
2 & 2 & 3 \\
1 & 1 & 2 \\
1 & 1 & 2
\end{pmatrix}
$$

we find

$$
\mathrm{tr} A = 2 = s_1, \quad \mathrm{tr} A^2 = 2 = s_2, \quad \mathrm{tr} A^3 = 5 = s_3 .
$$

We obtain the system of linear equations

$$
\begin{pmatrix}
1 & 0 & 0 \\
2 & 2 & 0 \\
2 & 2 & 3
\end{pmatrix}
\begin{pmatrix}
c_1 \\ c_2 \\ c_3
\end{pmatrix}
=
\begin{pmatrix}
-2 \\ -2 \\ -5
\end{pmatrix}
$$

with the solution

$$
c_1 = -2, \quad c_2 = 1, \quad c_3 = -1 .
$$

Since $c_3 = -1$ the inverse exists and $\det A = 1$. The inverse matrix of A is given by

$$
A^{-1} = \frac{1}{-c_3}(A^2 + c_1 A + c_2 I_3) = \begin{pmatrix}
0 & 1 & 0 \\
-1 & 1 & 1 \\
1 & -1 & 0
\end{pmatrix} .
$$

Problem 9.   Let U be the $n \times n$ unitary matrix

$$U := \begin{pmatrix} 0 & 1 & 0 & \cdots & 0 \\ 0 & 0 & 1 & \cdots & 0 \\ \vdots & \vdots & \vdots & \ddots & \vdots \\ 0 & 0 & 0 & \cdots & 1 \\ 1 & 0 & 0 & \cdots & 0 \end{pmatrix}$$

and V be the $n \times n$ unitary diagonal matrix ($\zeta \in \mathbf{C}$)

$$V := \begin{pmatrix} 1 & 0 & 0 & \cdots & 0 \\ 0 & \zeta & 0 & \cdots & 0 \\ 0 & 0 & \zeta^2 & \cdots & 0 \\ \vdots & \vdots & \vdots & \ddots & \vdots \\ 0 & 0 & 0 & \cdots & \zeta^{n-1} \end{pmatrix}$$

where $\zeta^n = 1$. Then the set of matrices

$$\{ U^j V^k \ : \ j, k = 0, 1, 2, \ldots, n-1 \}$$

provide a basis in the Hilbert space for all $n \times n$ matrices with the *scalar product*

$$\langle A, B \rangle := \frac{1}{n} \mathrm{tr}(AB^*)$$

for $n \times n$ matrices A and B. Write down the basis for $n = 2$.

Solution 9.   For $n = 2$ we have the combinations

$$(jk) \in \{ (00), (01), (10), (11) \} .$$

This yields the orthonormal basis in the vector space of 2×2 matrices

$$I_2 = \begin{pmatrix} 1 & 0 \\ 0 & 1 \end{pmatrix}, \quad \sigma_x = \begin{pmatrix} 0 & 1 \\ 1 & 0 \end{pmatrix}, \quad \sigma_z = \begin{pmatrix} 1 & 0 \\ 0 & -1 \end{pmatrix}, \quad -i\sigma_y = \begin{pmatrix} 0 & -1 \\ 1 & 0 \end{pmatrix} .$$

Problem 10.   Let A and B be $n \times n$ matrices over $\mathbf{C}$. Show that the matrices AB and BA have the same set of eigenvalues.

Solution 10.   Consider first the case that A is invertible. Then we have

$$AB = A(BA)A^{-1} .$$

Thus AB and BA are similar and therefore have the same set of eigenvalues. If A is singular we apply the *continuity argument*: If A is singular, consider $A + \epsilon I_n$. We choose $\delta > 0$ such that $A + \epsilon I_n$ is invertible for all ϵ, $0 < \epsilon < \delta$.

Thus $(A + \epsilon I_n)B$ and $B(A + \epsilon I_n)$ have the same set of eigenvalues for every $\epsilon \in (0, \delta)$. We equate their characteristic polynomials to obtain

$$\det(\lambda I_n - (A + \epsilon I_n)B) = \det(\lambda I_n - B(A + \epsilon I_n)), \quad 0 < \epsilon < \delta.$$

Since both sides are continuous (even analytic) functions of ϵ we find by letting $\epsilon \to 0^+$ that

$$\det(\lambda I_n - AB) = \det(\lambda I_n - BA).$$

Problem 11. An $n \times n$ *circulant matrix* C is given by

$$C := \begin{pmatrix} c_0 & c_1 & c_2 & \cdots & c_{n-1} \\ c_{n-1} & c_0 & c_1 & \cdots & c_{n-2} \\ c_{n-2} & c_{n-1} & c_0 & \cdots & c_{n-3} \\ \vdots & \vdots & \vdots & \ddots & \vdots \\ c_1 & c_2 & c_3 & \cdots & c_0 \end{pmatrix}.$$

For example, the matrix

$$P := \begin{pmatrix} 0 & 1 & 0 & \cdots & 0 \\ 0 & 0 & 1 & \cdots & 0 \\ \vdots & \vdots & \vdots & \ddots & \vdots \\ 0 & 0 & 0 & \cdots & 1 \\ 1 & 0 & 0 & \cdots & 0 \end{pmatrix}$$

is a circulant matrix. It is also called the $n \times n$ *primary permutation matrix*.
(i) Let C and P be the matrices given above. Let

$$f(\lambda) = c_0 + c_1 \lambda + \cdots + c_{n-1} \lambda^{n-1}.$$

Show that $C = f(P)$.
(ii) Show that C is a *normal matrix*, that is,

$$C^* C = C C^*.$$

(iii) Show that the eigenvalues of C are $f(\omega^k)$, $k = 0, 1, \ldots, n - 1$, where ω is the nth primitive root of unity.
(iv) Show that

$$\det(C) = f(\omega^0) f(\omega^1) \ldots f(\omega^{n-1}).$$

(v) Show that $F^* C F$ is a diagonal matrix, where F is the unitary matrix with (j, k)-entry equal to

$$\frac{1}{\sqrt{n}} \omega^{(j-1)(k-1)}, \quad j, k = 1, \ldots, n.$$

Solution 11.   (i) Direct calculation of

$$f(P) = c_0 I_n + c_1 P + c_2 P^2 + \cdots + c_{n-1} P^{n-1}$$

yields the matrix C, where I_n is the $n \times n$ unit matrix. Notice that P^2, $P^3, \ldots, P^{n-1}$ are permutation matrices.

(ii) We have $PP^* = P^*P$. If two $n \times n$ matrices A and B commute, then $g(A)$ and $h(B)$ commute, where g and h are polynomials. Thus C is a normal matrix.

(iii) The characteristic polynomial of P is

$$\det(\lambda I_n - P) = \lambda^n - 1 = \prod_{k=0}^{n-1} (\lambda - \omega^k).$$

Thus the eigenvalues of P and P^j are, respectively, ω^k and ω^{jk}, where $k = 0, 1, \ldots, n-1$. It follows that the eigenvalues of $C = f(P)$ are $f(\omega^k)$, $k = 0, 1, \ldots, n-1$.

(iv) Using the result from (iii) we find

$$\det(C) = \prod_{k=0}^{n-1} f(\omega^k).$$

(v) For each $k = 0, 1, \ldots, n-1$, let

$$\mathbf{x}_k = (1, \omega^k, \omega^{2k}, \ldots, \omega^{(n-1)k})^T$$

where T denotes the transpose. If follows that

$$P\mathbf{x}_k = (\omega^k, \omega^{2k}, \ldots, \omega^{(n-1)k}, 1)^T = \omega^k \mathbf{x}_k$$

and

$$C\mathbf{x}_k = f(P)\mathbf{x}_k = f(\omega^k)\mathbf{x}_k.$$

Thus the vectors $\mathbf{x}_k$ are the eigenvectors of P and C corresponding to the respective eigenvalues ω^k and $f(\omega^k)$, $k = 0, 1, \ldots, n-1$. Since

$$\langle \mathbf{x}_j, \mathbf{x}_k \rangle \equiv \mathbf{x}_j^* \mathbf{x}_k = \sum_{\ell=0}^{n-1} \overline{\omega^{k\ell}} \omega^{j\ell} = \sum_{\ell=0}^{n-1} \omega^{(j-k)\ell} = \begin{cases} 0 & j \neq k \\ n & j = k \end{cases}$$

we find that

$$\left\{ \frac{1}{\sqrt{n}} \mathbf{x}_0, \frac{1}{\sqrt{n}} \mathbf{x}_1, \ldots, \frac{1}{\sqrt{n}} \mathbf{x}_{n-1} \right\}$$

is an orthonormal basis in the Hilbert space $\mathbf{C}^n$. Thus we obtain the unitary matrix

$$F = \frac{1}{\sqrt{n}} \begin{pmatrix} 1 & 1 & 1 & \cdots & 1 \\ 1 & \omega & \omega^2 & \cdots & \omega^{n-1} \\ 1 & \omega^2 & \omega^4 & \cdots & \omega^{2(n-1)} \\ \vdots & \vdots & \vdots & \ddots & \vdots \\ 1 & \omega^{n-1} & \omega^{2(n-1)} & \cdots & \omega^{(n-1)(n-1)} \end{pmatrix}$$

such that
$$F^*CF = \mathrm{diag}(f(\omega^0), f(\omega^1), \ldots, f(\omega^{n-1})).$$

The matrix F is unitary and is called the *Fourier matrix*.

Problem 12. An $n \times n$ matrix A is called *reducible* if there is a permutation matrix P such that

$$P^T AP = \begin{pmatrix} B & C \\ 0 & D \end{pmatrix}$$

where B and D are square matrices of order at least 1. An $n \times n$ matrix A is called *irreducible* if it is not reducible. Show that the $n \times n$ primary permutation matrix

$$A := \begin{pmatrix} 0 & 1 & 0 & \cdots & 0 \\ 0 & 0 & 1 & \cdots & 0 \\ \vdots & \vdots & \vdots & \vdots & \vdots \\ 0 & 0 & 0 & \cdots & 1 \\ 1 & 0 & 0 & \cdots & 0 \end{pmatrix}$$

is irreducible.

Solution 12. Suppose the matrix A is reducible. Let

$$P^T AP = J_1 \oplus J_2 \oplus \ldots \oplus J_k, \quad k \geq 2$$

where P is some permutation matrix and the J_j are irreducible matrices of order $< n$. Here $\oplus$ denotes the direct sum. The rank of $A - I_n$ is $n - 1$ since $\det(A - I_n) = 0$ and the submatrix of size $n - 1$ by deleting the last row and the last column from $A - I_n$ is nonsingular. It follows that

$$\mathrm{rank}(P^T AP - I_n) = \mathrm{rank}(P^T(A - I_n)P) = n - 1.$$

By using the above decomposition, we obtain

$$\mathrm{rank}(P^T AP - I_n) = \sum_{j=1}^{k} \mathrm{rank}(J_j - I_n) \leq (n - k) < (n - 1).$$

This is a contradiction. Thus A is irreducible.

Problem 13. We define a linear *bijection*, h, between $\mathbf{R}^4$ and $\mathbf{H}(2)$, the set of complex 2×2 hermitian matrices, by

$$(t, x, y, z) \rightarrow \begin{pmatrix} t + x & y - iz \\ y + iz & t - x \end{pmatrix}.$$

We denote the matrix on the right-hand side by H.
(i) Show that the matrix can be written as a linear combination of the Pauli spin matrices and the identity matrix I_2.
(ii) Find the inverse map.
(iii) Calculate the determinant of 2×2 hermitian matrix H. Discuss.

Solution 13.   (i) We have

$$H = tI_2 + x\sigma_z + y\sigma_x + z\sigma_y .$$

(ii) Consider

$$\begin{pmatrix} a & c \\ c^* & b \end{pmatrix} = \begin{pmatrix} t+x & y-iz \\ y+iz & t-x \end{pmatrix} .$$

Compare the entries of the 2×2 matrix we obtain

$$t = \frac{a+b}{2}, \quad x = \frac{a-b}{2}, \quad y = \frac{c+c^*}{2}, \quad z = \frac{c^*-c}{2i} .$$

(iii) We obtain

$$\det H = t^2 - x^2 - y^2 - z^2 .$$

This is the *Lorentz metric.* Let U be a unitary 2×2 matrix. Then $\det(UHU^*) = \det(H)$.

Problem 14.   Let A be an $n \times n$ invertible matrix over $\mathbf{C}$. Assume that A can be written as

$$A = B + iB$$

where B has only real coefficients. Show that B^{-1} exists and

$$A^{-1} = \frac{1}{2}(B^{-1} - iB^{-1}) .$$

Solution 14.   Since

$$A = (1+i)B$$

and $\det A \neq 0$ we have $(1+i)^n \det B \neq 0$. Thus $\det B \neq 0$ and B^{-1} exists. We have

$$(1+i)B\frac{1}{2}(1-i)B^{-1} = I_n .$$

Problem 15.   Let A be an invertible matrix. Assume that $A = A^{-1}$. What are the possible values for $\det A$?

Solution 15. Since

$$1 = \det I_n = \det(AA^{-1}) = \det A \det A^{-1}$$

and by assumption $\det A = \det A^{-1}$ we have

$$1 = (\det A)^2 .$$

Thus $\det A$ is either $+1$ or -1.

Problem 16. Let A be a skew-symmetric matrix over $\mathbf{R}$, i.e. $A^T = -A$ and of order $2n - 1$. Show that $\det(A) = 0$.

Solution 16. From $A^T = -A$ we obtain

$$\det(A^T) = \det(-A) = (-1)^{2n-1} \det(A) = - \det(A) .$$

Since

$$\det(A) = \det(A^T)$$

we obtain $\det(A) = - \det(A)$ and therefore $\det(A) = 0$.

Problem 17. Show that if A is hermitian, i.e. $A^* = A$ then $\det(A)$ is a real number.

Solution 17. Since A is hermitian we have $A^* = A$ or $\overline{A} = A^T$. Furthermore $\det(A) = \det(A^T)$. Thus

$$\det(\overline{A}) = \det(A^T) = \det A .$$

Now if $\det(A) = x + iy$ then $\det(\overline{A}) = x - iy$ with $x, y \in \mathbf{R}$. Thus $x + iy = x - iy$ and therefore $y = 0$. Thus $\det(A)$ is a real number.

Problem 18. Let A, B, and C be $n \times n$ matrices. Calculate

$$\det \begin{pmatrix} A & 0_n \\ C & B \end{pmatrix} .$$

where 0_n is the $n \times n$ zero matrix.

Solution 18. Obviously we find

$$\det \begin{pmatrix} A & 0_n \\ C & B \end{pmatrix} = \det(A) \det(B) .$$

Problem 19. Let A, B are 2×2 matrices over $\mathbf{R}$. Let $H := A + iB$. Express $\det H$ as a sum of determinants.

Solution 19.   Since

$$H = \begin{pmatrix} a_{11} + ib_{11} & a_{12} + ib_{12} \\ a_{21} + ib_{21} & b_{22} + ib_{22} \end{pmatrix}$$

we find

$$\det H = a_{11}a_{22} - a_{12}a_{21} - b_{11}b_{22} + b_{12}b_{21}i(a_{11}b_{22} + b_{11}a_{22} - a_{21}b_{12} - a_{12}b_{21})$$

$$= \det A - \det B + i \det \begin{pmatrix} a_{11} & a_{12} \\ b_{21} & b_{22} \end{pmatrix} + i \det \begin{pmatrix} b_{11} & b_{12} \\ a_{21} & a_{22} \end{pmatrix}.$$

Problem 20.   Let A, B are 2×2 matrices over $\mathbf{R}$. Let $H := A + iB$. Assume that H is hermitian. Show that

$$\det H = \det A - \det B.$$

Solution 20.   Since H is hermitian, i.e. $H^* = H$ with $H^* = A^T - iB^T$ we have $A = A^T$ and $B = -B^T$. Thus $a_{12} = a_{21}$, $b_{11} = b_{22} = 0$ and $b_{12} = -b_{21}$. It follows that

$$\det \begin{pmatrix} a_{11} + ib_{11} & a_{12} + ib_{12} \\ a_{21} + ib_{21} & a_{22} + ib_{22} \end{pmatrix} = \det A - \det B$$

$$+ i(a_{11}b_{22} + b_{11}a_{22} - a_{21}b_{12} - a_{12}b_{21})$$

$$= \det A - \det B.$$

Problem 21.   Let A, B, C, D be $n \times n$ matrices. Assume that $DC = CD$, i.e. C and D commute and $\det D \neq 0$. Consider the $(2n) \times (2n)$ matrix

$$M = \begin{pmatrix} A & B \\ C & D \end{pmatrix}.$$

Show that

$$\det M = \det(AD - BC). \tag{1}$$

We know that

$$\det \begin{pmatrix} U & 0_n \\ X & Y \end{pmatrix} = \det U \det Y \tag{2}$$

and

$$\det \begin{pmatrix} U & V \\ 0_n & Y \end{pmatrix} = \det U \det Y \tag{3}$$

where U, V, X, Y are $n \times n$ matrices and 0_n is the $n \times n$ zero matrix.

Solution 21. We have the identity

$$\begin{pmatrix} A & B \\ C & D \end{pmatrix}\begin{pmatrix} D & 0_n \\ -C & I_n \end{pmatrix} = \begin{pmatrix} AD - BC & B \\ CD - DC & D \end{pmatrix} = \begin{pmatrix} AD - BC & B \\ 0_n & D \end{pmatrix} \quad (4)$$

where we used that $CD = DC$. Applying the determinant to the right and left-hand side of (4) and using (2),(3) and $\det D \neq 0$ we obtain identity (1).

Problem 22. Let A, B be $n \times n$ matrices. We have the identity

$$\det \begin{pmatrix} A & B \\ B & A \end{pmatrix} \equiv \det(A + B)\det(A - B).$$

Use this identity to calculate the determinant of the left-hand side using the right-hand side, where

$$A = \begin{pmatrix} 2 & 3 \\ 1 & 7 \end{pmatrix}, \qquad B = \begin{pmatrix} 0 & 2 \\ 4 & 6 \end{pmatrix}.$$

Solution 22. We have

$$A + B = \begin{pmatrix} 2 & 5 \\ 5 & 13 \end{pmatrix}, \qquad A - B = \begin{pmatrix} 2 & 1 \\ -3 & 1 \end{pmatrix}.$$

Therefore $\det(A + B) = 1$ and $\det(A - B) = 5$. Finally,

$$\det(A + B)\det(A - B) = 5.$$

Problem 23. Let A, B, C, D be $n \times n$ matrices. Assume that D is invertible. Consider the $(2n) \times (2n)$ matrix

$$M = \begin{pmatrix} A & B \\ C & D \end{pmatrix}.$$

Show that

$$\det M = \det(AD - BD^{-1}CD). \quad (1)$$

Solution 23. Using the identity

$$\begin{pmatrix} A & B \\ C & D \end{pmatrix}\begin{pmatrix} I_n & 0_n \\ -D^{-1}C & I_n \end{pmatrix} \equiv \begin{pmatrix} A - BD^{-1}C & B \\ 0_n & D \end{pmatrix}$$

equation (1) follows.

Problem 24.   Let A, B be $n \times n$ positive definite (and therefore hermitian) matrices. Show that

$$\mathrm{tr}(AB) > 0.$$

Solution 24.   Let U be a unitary matrix such that

$$U^* AU = D = \mathrm{diag}(d_1, d_2, \ldots, d_n).$$

Obviously, d_1, d_2, $\ldots$, d_n are the eigenvalues of A. Then $\mathrm{tr}A = \mathrm{tr}D$ and

$$\mathrm{tr}(AB) = \mathrm{tr}(U^* AUU^* BU) = \mathrm{tr}(DC)$$

where $C = U^* BU$. Now C is positive definite and therefore its diagonal entries c_{ii} are real and positive and $\mathrm{tr}C = \mathrm{tr}B$. The diagonal entries of DC are $d_i c_{ii}$ and therefore

$$\mathrm{tr}(DC) = \sum_{i=1}^{n} d_i c_{ii} > 0.$$

Problem 25.   Let $P_0(x) = 1$, $P_1(x) = \alpha_1 - x$ and

$$P_k(x) = (\alpha_k - x)P_{k-1}(x) - \beta_{k-1}P_{k-2}(x), \qquad k = 2, 3, \ldots$$

where β_j, $j = 1, 2, \ldots$ are positive numbers. Find a $k \times k$ matrix A_k such that

$$P_k(x) = \det(A_k).$$

Solution 25.   The matrix is

$$A_k = \begin{pmatrix} \alpha_1 - x & \beta_1 & 0 & \cdots & 0 \\ 1 & \alpha_2 - x & \beta_2 & \cdots & 0 \\ 0 & 1 & \alpha_3 - x & \cdots & 0 \\ \vdots & \vdots & \vdots & \ddots & \vdots \\ 0 & \cdots & 0 & 1 & \alpha_k - x \end{pmatrix}.$$

Problem 26.   Let

$$A = \begin{pmatrix} \dfrac{1}{x_1 + y_1} & \dfrac{1}{x_1 + y_2} \\ \dfrac{1}{x_2 + y_1} & \dfrac{1}{x_2 + y_2} \end{pmatrix}$$

where we assume that $x_i + y_j \neq 0$ for $i, j = 1, 2$. Show that

$$\det A = \frac{(x_1 - x_2)(y_1 - y_2)}{(x_1 + y_1)(x_1 + y_2)(x_2 + y_1)(x_2 + y_2)}.$$

Solution 26. We have

$$\begin{aligned}
\det A &= \frac{1}{x_1 + y_1} \frac{1}{x_2 + y_2} - \frac{1}{x_2 + y_1} \frac{1}{x_1 + y_2} \\
&= \frac{(x_2 + y_1)(x_1 + y_2) - (x_1 + y_1)(x_2 + y_2)}{(x_1 + y_1)(x_1 + y_2)(x_2 + y_1)(x_2 + y_2)} \\
&= \frac{x_1 y_1 - x_1 y_2 + x_2 y_2 - x_2 y_1}{(x_1 + y_1)(x_1 + y_2)(x_2 + y_1)(x_2 + y_2)} \\
&= \frac{(x_1 - x_2)(y_1 - y_2)}{(x_1 + y_1)(x_1 + y_2)(x_2 + y_1)(x_2 + y_2)}.
\end{aligned}$$

For the general case with the matrix

$$A = \begin{pmatrix}
\dfrac{1}{x_1 + y_1} & \dfrac{1}{x_1 + y_2} & \dfrac{1}{x_1 + y_3} & \cdots & \dfrac{1}{x_1 + y_n} \\
\dfrac{1}{x_2 + y_1} & \dfrac{1}{x_2 + y_2} & \dfrac{1}{x_2 + y_3} & \cdots & \dfrac{1}{x_2 + y_n} \\
\vdots & \vdots & \vdots & \ddots & \vdots \\
\dfrac{1}{x_n + y_1} & \dfrac{1}{x_n + y_2} & \dfrac{1}{x_n + y_3} & \cdots & \dfrac{1}{x_n + y_n}
\end{pmatrix}$$

we find the determinant (*Cauchy determinant*)

$$\det A = \frac{\prod_{i>j}^{n}(x_i - x_j)(y_i - y_j)}{\prod_{i,j=1}^{n}(x_i + y_j)}.$$

Problem 27. For a 3×3 matrix we can use the *rule of Sarrus* to calculate the determinant (for higher dimensions there is no such thing). Let

$$\begin{pmatrix}
a_{11} & a_{12} & a_{13} \\
a_{21} & a_{22} & a_{23} \\
a_{31} & a_{32} & a_{33}
\end{pmatrix}.$$

Write the first two columns again to the right of the matrix to obtain

$$\begin{pmatrix}
a_{11} & a_{12} & a_{13} & | & a_{11} & a_{12} \\
a_{21} & a_{22} & a_{23} & | & a_{21} & a_{22} \\
a_{31} & a_{32} & a_{33} & | & a_{31} & a_{32}
\end{pmatrix}.$$

Now look at the diagonals. The product of the diagonals sloping down to the right have a plus sign, the ones up to the left have a negative sign. This leads to the determinant

$$\det A = a_{11}a_{22}a_{33}+a_{12}a_{23}a_{31}+a_{13}a_{21}a_{32}-a_{31}a_{22}a_{13}-a_{32}a_{23}a_{11}-a_{33}a_{21}a_{12}.$$

Use this rule to calculate the determinant of the *rotational matrix*

$$R = \begin{pmatrix} \cos\theta & 0 & -\sin\theta \\ 0 & 1 & 0 \\ \sin\theta & 0 & \cos\theta \end{pmatrix}.$$

Solution 27.   We have

$$\begin{pmatrix} \cos\theta & 0 & -\sin\theta & | & \cos\theta & 0 \\ 0 & 1 & 0 & | & 0 & 1 \\ \sin\theta & 0 & \cos\theta & | & \sin\theta & 0 \end{pmatrix}.$$

Thus

$$\det R = \cos\theta \cdot \cos\theta + 0 + 0 - \sin\theta \cdot (-\sin\theta) - 0 - 0 = \cos^2\theta + \sin^2\theta = 1.$$

Problem 28.   Let A, S be $n \times n$ matrices. Assume that S is invertible and assume that

$$S^{-1}AS = \rho S$$

where $\rho \neq 0$. Show that A is invertible.

Solution 28.   From $S^{-1}AS = \rho S$ we obtain

$$\det(S^{-1}AS) = \det(\rho S).$$

Thus

$$\det(S^{-1})\det(A)\det(S) = \det(\rho S).$$

It follows that

$$\det(A) = \det(\rho S) = \rho^n \det(S).$$

Since $\rho^n \neq 0$ and $\det S \neq 0$ it follows that $\det A \neq 0$ and therefore A^{-1} exists.

Problem 29.   The determinant of an $n \times n$ *circulant matrix* is given by

$$\det \begin{pmatrix} a_1 & a_2 & a_3 & \cdots & a_n \\ a_n & a_1 & a_2 & \cdots & a_{n-1} \\ \vdots & \vdots & \vdots & \ddots & \vdots \\ a_3 & a_4 & a_5 & \cdots & a_2 \\ a_2 & a_3 & a_4 & \cdots & a_1 \end{pmatrix} = (-1)^{n-1}\prod_{j=0}^{n-1}\left(\sum_{k=1}^{n}\zeta^{jk}a_k\right) \qquad (1)$$

where $\zeta := \exp(2\pi i/n)$. Find the determinant of the circulant $n \times n$ matrix

$$\begin{pmatrix} 1 & 4 & 9 & \cdots & n^2 \\ n^2 & 1 & 4 & \cdots & (n-1)^2 \\ \vdots & \vdots & \vdots & \ddots & \vdots \\ 9 & 16 & 25 & \cdots & 4 \\ 4 & 9 & 16 & \cdots & 1 \end{pmatrix}$$

using equation (1).

Solution 29. Applying (1) we have

$$\det\begin{pmatrix} 1 & 4 & 9 & \cdots & n^2 \\ n^2 & 1 & 4 & \cdots & (n-1)^2 \\ \vdots & \vdots & \vdots & \ddots & \vdots \\ 9 & 16 & 25 & \cdots & 4 \\ 4 & 9 & 16 & \cdots & 1 \end{pmatrix} = (-1)^{n-1} \prod_{j=0}^{n-1}\left(\sum_{k=1}^{n} \zeta^{jk} k^2\right).$$

We have

$$\sum_{k=1}^{n} k^2 x^k = \frac{n^2 x^{n+3} - (2n^2 + 2n - 1)x^{n+2} + (n^2 + 2n + 1)x^{n+1} - x^2 - x}{(x-1)^3}$$

for $x \neq 1$ and

$$\sum_{k=1}^{n} k^2 x^k = \frac{n(n+1)(2n+1)}{6}$$

for $x = 1$. For the product we have

$$\prod_{j=1}^{n-1}(\zeta^j - a) = (-1)^{n-1}\sum_{k=0}^{n-1} a_k$$

$$= \begin{cases} (-1)^{n-1}\frac{a^n - 1}{a-1} & \text{if } a \neq 1 \\ (-1)^{n-1}n & \text{if } a = 1. \end{cases}$$

It follows that

$$\prod_{j=1}^{n-1} \zeta^j = (-1)^{n-1}, \qquad \prod_{j=1}^{n-1}(\zeta^j - 1) = (-1)^{n-1}n$$

and

$$\prod_{j=1}^{n-1}\left(\zeta^j - \frac{n+2}{n}\right) = (-1)^{n-1}\frac{(n+2)^n - n^n}{2n^{n-1}}.$$

Finally

$$\det \begin{pmatrix} 1 & 4 & 9 & \cdots & n^2 \\ n^2 & 1 & 4 & \cdots & (n-1)^2 \\ \vdots & \vdots & \vdots & \ddots & \vdots \\ 9 & 16 & 25 & \cdots & 4 \\ 4 & 9 & 16 & \cdots & 1 \end{pmatrix}$$

$$= (-1)^{n-1} \frac{n^{n-2}(n+1)(2n+1)((n+2)^n - n^n)}{12}.$$

Problem 30. Let A be a nonzero 2×2 matrix over $\mathbf{R}$. Let B_1, B_2, B_3, B_4 be 2×2 matrices over $\mathbf{R}$ and assume that

$$\det(A + B_j) = \det A + \det B_j \qquad \text{for} \quad j = 1, 2, 3, 4.$$

Show that there exist real numbers c_1, c_2, c_3, c_4, not all zero, such that

$$c_1 B_1 + c_2 B_2 + c_3 B_3 + c_4 B_4 = \begin{pmatrix} 0 & 0 \\ 0 & 0 \end{pmatrix}. \tag{1}$$

Solution 30. Let

$$A = \begin{pmatrix} a_{11} & a_{12} \\ a_{21} & a_{22} \end{pmatrix}, \qquad B_j = \begin{pmatrix} b_{11}^{(j)} & b_{12}^{(j)} \\ b_{21}^{(j)} & b_{22}^{(j)} \end{pmatrix}$$

with $j = 1, 2, 3, 4$. If $\det(A + B_j) = \det A + \det B_j$, it follows that

$$a_{22} b_{11}^{(j)} - a_{21} b_{12}^{(j)} - a_{12} b_{21}^{(j)} + a_{11} b_{22}^{(j)} = 0. \tag{2}$$

Since A is a nonzero matrix the solution space to (1) for fixed j is a three-dimensional vector space. Any four vectors in a three-dimensional space are linearly dependent. Thus there must exist c_1, c_2, c_3, c_4, not all 0, for which equation (1) is true.

Problem 31. Let A, B be $n \times n$ matrices. Show that

$$\text{tr}((A+B)(A-B)) = \text{tr}A^2 - \text{tr}B^2. \tag{1}$$

Solution 31. We have

$$(A+B)(A-B) = A^2 - AB + BA - B^2 = A^2 + [B, A] - B^2.$$

Since $\text{tr}([B, A]) = 0$ and the trace is linear we obtain (1).

Problem 32. An $n \times n$ matrix Q is *orthogonal* if Q is real and

$$Q^T Q = Q^T Q = I_n$$

i.e.

$$Q^{-1} = Q^T.$$

(i) Find the determinant of an orthogonal matrix.
(ii) Let $\mathbf{u}$, $\mathbf{v}$ be two vectors in $\mathbf{R}^3$ and $\mathbf{u} \times \mathbf{v}$ denotes the *vector product* of $\mathbf{u}$ and $\mathbf{v}$

$$\mathbf{u} \times \mathbf{v} := \begin{pmatrix} u_2 v_3 - u_3 v_2 \\ u_3 v_1 - u_1 v_3 \\ u_1 v_2 - u_2 v_1 \end{pmatrix}.$$

Let Q be a 3×3 orthogonal matrix. Calculate

$$(Q\mathbf{u}) \times (Q\mathbf{v}).$$

Solution 32. (i) We have

$$\det(Q) = \det(Q^T) = \det(Q^{-1}) = 1/\det(Q).$$

Thus $(\det(Q))^2 = 1$ and therefore

$$\det(Q) = \pm 1.$$

(ii) We have

$$(Q\mathbf{u}) \times (Q\mathbf{v}) = Q(\mathbf{u} \times \mathbf{v}) \det(Q).$$

Problem 33. Calculate the determinant of the $n \times n$ matrix

$$A = \begin{pmatrix} 1 & 1 & 1 & 1 & \cdots & 1 & 1 \\ 1 & 0 & 1 & 1 & \cdots & 1 & 1 \\ 1 & 1 & 0 & 1 & \cdots & 1 & 1 \\ 1 & 1 & 1 & 0 & \cdots & 1 & 1 \\ \vdots & \vdots & \vdots & \vdots & \ddots & \vdots & \vdots \\ 1 & 1 & 1 & 1 & \cdots & 0 & 1 \\ 1 & 1 & 1 & 1 & \cdots & 1 & 0 \end{pmatrix}.$$

Solution 33. Subtracting the second row from the first row ($n \geq 2$) we find the matrix

$$B = \begin{pmatrix} 0 & 1 & 0 & 0 & \cdots & 0 & 0 \\ 1 & 0 & 1 & 1 & \cdots & 1 & 1 \\ 1 & 1 & 0 & 1 & \cdots & 1 & 1 \\ 1 & 1 & 1 & 0 & \cdots & 1 & 1 \\ \vdots & \vdots & \vdots & \vdots & \ddots & \vdots & \vdots \\ 1 & 1 & 1 & 1 & \cdots & 0 & 1 \\ 1 & 1 & 1 & 1 & \cdots & 1 & 0 \end{pmatrix}.$$

Thus $\det A = \det B$. Let $s_n := \det B$. Now an expansion yields

$$s_n = -s_{n-1}$$

with the initial value $s_1 = 1$ from the matrix A. It follows that

$$s_n = \det A = (-1)^{n-1}, \qquad n = 1, 2, \ldots .$$

Problem 34.   Find the determinant of the matrix

$$A = \begin{pmatrix} 0 & 1 & 1 & 1 & \cdots & 1 & 1 \\ 1 & 0 & 1 & 1 & \cdots & 1 & 1 \\ 1 & 1 & 0 & 1 & \cdots & 1 & 1 \\ 1 & 1 & 1 & 0 & \cdots & 1 & 1 \\ \vdots & \vdots & \vdots & \vdots & \ddots & \vdots & \vdots \\ 1 & 1 & 1 & 1 & \cdots & 0 & 1 \\ 1 & 1 & 1 & 1 & \cdots & 1 & 0 \end{pmatrix} .$$

Solution 34.   Using the result from the previous problem we find

$$\det A = (-1)^{n-1}(n-1) .$$

Problem 35.   Let A be a 2×2 matrix over $\mathbf{R}$

$$A = \begin{pmatrix} a_{11} & a_{12} \\ a_{21} & a_{22} \end{pmatrix}$$

with $\det A \neq 0$. Is $(A^T)^{-1} = (A^{-1})^T$?

Solution 35.   We have

$$A^{-1} = \frac{1}{\det A} \begin{pmatrix} a_{22} & -a_{12} \\ -a_{21} & a_{11} \end{pmatrix}, \qquad (A^{-1})^T = \frac{1}{\det A} \begin{pmatrix} a_{22} & -a_{21} \\ -a_{12} & a_{11} \end{pmatrix}$$

and

$$A^T = \begin{pmatrix} a_{11} & a_{21} \\ a_{12} & a_{22} \end{pmatrix}, \qquad (A^T)^{-1} = \frac{1}{\det A} \begin{pmatrix} a_{22} & -a_{21} \\ -a_{12} & a_{11} \end{pmatrix} .$$

Thus $(A^T)^{-1} = (A^{-1})^T$. This is true also for $n \times n$ matrices.

Problem 36.   Let A be an invertible $n \times n$ matrix. Let $c = 2$. Can we find an invertible matrix S such that

$$SAS^{-1} = cA .$$

Solution 36. From $SAS^{-1} = cA$ it follows that

$$\det(SAS^{-1}) = \det(cA)$$
$$\det A = c^n \det A.$$

Since $\det A \neq 0$ we have $c^n = 1$. Thus such an S cannot be found.

Problem 37. Let

$$A = \begin{pmatrix} a_1 & a_2 \\ a_2 & a_3 \end{pmatrix}, \qquad B = \begin{pmatrix} b_1 & b_2 \\ b_2 & b_3 \end{pmatrix}$$

be symmetric matrices over **R**. The *Hadamard product* $A \bullet B$ is defined as

$$A \bullet B := \begin{pmatrix} a_1 b_1 & a_2 b_2 \\ a_2 b_2 & a_3 b_3 \end{pmatrix}.$$

Assume that A and B are positive definite. Show that $A \bullet B$ is positive definite using the trace and determinant.

Solution 37. We have

$$A \quad \text{positive definite} \quad \Leftrightarrow \quad \text{tr} A > 0 \quad \det A > 0$$
$$B \quad \text{positive definite} \quad \Leftrightarrow \quad \text{tr} B > 0 \quad \det B > 0.$$

Thus we have $a_1 + a_3 > 0$, $a_1 a_3 - a_2^2 > 0$ (or $a_1 a_3 > a_2^2$) and $b_1 + b_3 > 0$, $b_1 b_3 - b_2^2 > 0$ (or $b_1 b_3 > b_2^2$). Now using the conditions $a_1 a_3 > a_2^2$ and $b_1 b_3 > b_2^2$ we have

$$\det(A \bullet B) = (a_1 a_3)(b_1 b_3) - (a_2^2)(b_2^2)$$
$$> (a_2^2)(b_2^2) - (a_2^2)(b_2^2)$$
$$= 0.$$

Thus $\det(A \bullet B) > 0$. For the trace we have

$$\text{tr}(A \bullet B) = a_1 b_1 + a_3 b_3.$$

Now using $a_1 a_3 > a_2^2$ and $b_1 b_3 > b_2^2$ again we have

$$(a_1 + a_3)(a_1 b_1 + a_3 b_3)(b_1 + b_3) = a_1^2 b_1^2 + a_1 a_3 b_1^2 + a_1 a_3 b_1 b_3 + a_3^2 b_1 b_3$$
$$+ a_1^2 b_1 b_3 + a_1 a_3 b_1 b_3 + a_1 a_3 b_3^2 + a_3^2 b_3^2$$
$$> a_1^2 b_1^2 + a_2^2 b_1^2 + a_2^2 b_2^2 + a_3^2 b_2^2$$
$$+ a_1^2 b_2^2 + a_2^2 b_2^2 + a_2^2 b_3^2 + a_3^2 b_3^2$$
$$> 0.$$

Since $a_1 + a_3 > 0$ and $b_1 + b_3 > 0$ it follows that $a_1 b_1 + a_3 b_3 > 0$ and therefore $\text{tr}(A \bullet B) > 0$. Thus since $\det(A \bullet B) > 0$ and $\text{tr}(A \bullet B) > 0$ it follows that $A \bullet B$ is positive definite.

Chapter 4

Eigenvalues and Eigenvectors

Let A be an $n \times n$ matrix over $\mathbf{C}$. Then the complex number λ is called an *eigenvalue* of A if and only if the matrix $(A - \lambda I_n)$ is singular. Let $\mathbf{x}$ be a nonzero column vector in $\mathbf{C}^n$. Then $\mathbf{x}$ is called an *eigenvector* belonging to (or associated with) the eigenvalue λ if and only if $(A - \lambda I_n)\mathbf{x} = 0$. The equation

$$A\mathbf{x} = \lambda\mathbf{x}$$

is called the *eigenvalue equation*. For an eigenvector $\mathbf{x}$ of A, $A\mathbf{x}$ is a scalar multiple of $\mathbf{x}$. From the eigenvalue equation we obtain

$$\mathbf{x}^* A^* = \lambda^* \mathbf{x}^*$$

and thus

$$\mathbf{x}^* A^* A \mathbf{x} = \lambda\lambda^* \mathbf{x}^* \mathbf{x}.$$

If the matrix is hermitian then the eigenvalues are real.
If the matrix is unitary then the absolute value of the eigenvalues is 1.
If the matrix is skew-hermitian then the eigenvalues are purely imaginary.
If the matrix is a projection matrix then the eigenvalues can only take the values 1 and 0.
If the matrix is a permutation matrix then $\lambda_j \in \{1, -1\}$.
The eigenvalues of an upper or lower triangular matrix are the elements on the diagonal.

Problem 1. (i) Find the eigenvalues and normalized eigenvectors of the rotational matrix

$$A = \begin{pmatrix} \sin\theta & \cos\theta \\ -\cos\theta & \sin\theta \end{pmatrix}.$$

(ii) Are the eigenvectors orthogonal to each other?

Solution 1. (i) From the characteristic equation

$$\lambda^2 - 2\lambda\sin\theta + 1 = 0$$

we obtain the eigenvalues

$$\lambda_1 = \sin\theta + i\cos\theta, \qquad \lambda_2 = \sin\theta - i\cos\theta.$$

The corresponding normalized eigenvectors are

$$\mathbf{x}_1 = \frac{1}{\sqrt{2}}\begin{pmatrix} -i \\ 1 \end{pmatrix}, \qquad \mathbf{x}_2 = \frac{1}{\sqrt{2}}\begin{pmatrix} i \\ 1 \end{pmatrix}.$$

(ii) We see that $\mathbf{x}_1^*\mathbf{x}_2 = 0$. Thus the eigenvectors are orthogonal to each other.

Problem 2. (i) An $n \times n$ matrix A such that $A^2 = A$ is called *idempotent*. What can be said about the eigenvalues of such a matrix?
(ii) An $n \times n$ matrix A for which $A^p = 0_n$, where p is a positive integer, is called *nilpotent*. What can be said about the eigenvalues of such a matrix?
(iii) An $n \times n$ matrix A such that $A^2 = I_n$ is called *involutory*. What can be said about the eigenvalues of such a matrix?

Solution 2. (i) From the eigenvalue equation $A\mathbf{x} = \lambda\mathbf{x}$ we obtain

$$A^2\mathbf{x} = A(A\mathbf{x}) = A\lambda\mathbf{x} = \lambda A\mathbf{x} = \lambda^2\mathbf{x}.$$

Thus since $A^2 = A$ we have $\lambda^2 = \lambda$.
(ii) From the eigenvalue equation $A\mathbf{x} = \lambda\mathbf{x}$ we obtain $A^p\mathbf{x} = \lambda^p\mathbf{x}$. Since $A^p = 0_n$ we obtain $\lambda^p = 0$. Thus all eigenvalues are 0.
(iii) From the eigenvalue equation $A\mathbf{x} = \lambda\mathbf{x}$ we obtain $A^2\mathbf{x} = \lambda^2\mathbf{x}$. Since $A^2 = I_n$ we have $\lambda^2 = 1$. Thus $\lambda = \pm 1$.

Problem 3. Let $\mathbf{x}$ be a nonzero column vector in $\mathbf{R}^n$. Then $\mathbf{x}\mathbf{x}^T$ is an $n \times n$ matrix and $\mathbf{x}^T\mathbf{x}$ is a real number. Show that $\mathbf{x}^T\mathbf{x}$ is an eigenvalue of $\mathbf{x}\mathbf{x}^T$ and $\mathbf{x}$ is the corresponding eigenvector.

Solution 3. Since matrix multiplication is associative we have

$$(\mathbf{x}\mathbf{x}^T)\mathbf{x} = \mathbf{x}(\mathbf{x}^T\mathbf{x}) = (\mathbf{x}^T\mathbf{x})\mathbf{x}.$$

Problem 4.   Let A be an $n \times n$ matrix over $\mathbf{C}$. Show that the eigenvectors corresponding to distinct eigenvalues are linearly independent.

Solution 4.   The proof is by contradiction. Consider the eigenvalue equations

$$Ax = \lambda_1 x, \qquad Ay = \lambda_2 y, \qquad x, y \neq 0$$

with $\lambda_1 \neq \lambda_2$. Assume that $x = cy$ with $c \neq 0$, i.e., we assume that the eigenvectors x and y are linearly dependent. Then from $Ax = \lambda_1 x$ we obtain $A(cy) = \lambda_1(cy)$. Thus

$$c(Ay) = c(\lambda_1 y).$$

Using $Ay = \lambda_2 y$ we arrive at

$$c(\lambda_1 - \lambda_2)y = 0.$$

Since $c \neq 0$ and $y \neq 0$ we obtain $\lambda_1 = \lambda_2$. Thus we have a contradiction and therefore x and y must be linearly independent.

Problem 5.   Let A be an $n \times n$ matrix over $\mathbf{C}$. The *spectral radius* of the matrix A is the non-negative number defined by

$$\rho(A) := \max\{ |\lambda_j(A)| : 1 \leq j \leq n \}$$

where $\lambda_j(A)$ are the eigenvalues of A. We define the *norm* of A as

$$\|A\| := \sup_{\|x\|=1} \|Ax\|$$

where $\|Ax\|$ denotes the Euclidean norm of the vector Ax. Show that $\rho(A) \leq \|A\|$.

Solution 5.   From the eigenvalue equation $Ax = \lambda x$ ($x \neq 0$) we obtain

$$|\lambda|\|x\| = \|\lambda x\| = \|Ax\| \leq \|A\|\|x\|.$$

Therefore since $\|x\| \neq 0$ we obtain $|\lambda| \leq \|A\|$ and $\rho(A) \leq \|A\|$.

Problem 6.   Let A be an $n \times n$ hermitian matrix, i.e., $A = A^*$. Assume that all n eigenvalues are different. Then the normalized eigenvectors $\{ v_j : j = 1, 2, \ldots, n \}$ form an orthonormal basis in $\mathbf{C}^n$. Consider

$$\beta := (Ax - \mu x, Ax - \nu x) \equiv (Ax - \mu x)^*(Ax - \nu x)$$

where $(\,,\,)$ denotes the scalar product in $\mathbf{C}^n$ and μ, ν are real constants with $\mu < \nu$. Show that if no eigenvalue lies between μ and ν, then $\beta \geq 0$.

Solution 6. First note that $\lambda_j \in \mathbf{R}$, since the matrix A is hermitian. We expand $\mathbf{x}$ with respect to the basis $\{ \mathbf{v}_j : j = 1, 2, \ldots, n \}$, i.e.

$$\mathbf{x} = \sum_{j=1}^{n} \alpha_j \mathbf{v}_j .$$

Since $A\mathbf{v}_j = \lambda_j \mathbf{v}_j$ and therefore $\mathbf{v}_j^* A = \lambda_j \mathbf{v}_j^*$ we have

$$A\mathbf{x} = \sum_{j=1}^{n} \alpha_j \lambda_j \mathbf{v}_j, \qquad \mathbf{x}^* A = \sum_{k=1}^{n} \alpha_k^* \lambda_k \mathbf{v}_k^* .$$

Since $(\mathbf{v}_j, \mathbf{v}_k) = \mathbf{v}_j^* \mathbf{v}_k = \delta_{jk}$ we find

$$\beta = \mathbf{x}^* A A \mathbf{x} - \nu \mathbf{x}^* A \mathbf{x} - \mu \mathbf{x}^* A \mathbf{x} + \mu\nu \mathbf{x}^* \mathbf{x}$$

$$= \sum_{j=1}^{n} \lambda_j^2 |\alpha_j|^2 - \nu \sum_{j=1}^{n} \lambda_j |\alpha_j|^2 - \mu \sum_{j=1}^{n} \lambda_j |\alpha_j|^2 + \mu\nu \sum_{j=1}^{n} |\alpha_j|^2$$

$$= \sum_{j=1}^{n} (\lambda_j - \mu)(\lambda_j - \nu) |\alpha_j|^2 .$$

If no eigenvalue lies between μ and ν, then $(\lambda_j - \mu)$ and $(\lambda_j - \nu)$ have the same sign for all eigenvalues. Therefore $(\lambda_j - \mu)(\lambda_j - \nu)$ is nonnegative. Thus $\beta \geq 0$.

Problem 7. Let Q and P be $n \times n$ symmetric matrices over $\mathbf{R}$, i.e., $Q = Q^T$ and $P = P^T$. Assume that P^{-1} exists. Find the maximum of the function $f : \mathbf{R}^n \to \mathbf{R}$

$$f(\mathbf{x}) = \mathbf{x}^T Q \mathbf{x}$$

subject to $\mathbf{x}^T P \mathbf{x} = 1$. Use the *Lagrange multiplier method*.

Solution 7. The Lagrange function is

$$L(\mathbf{x}, \lambda) = \mathbf{x}^T Q \mathbf{x} + \lambda(1 - \mathbf{x}^T P \mathbf{x})$$

where λ is the Lagrange multiplier. The partial derivatives lead to

$$\frac{\partial L}{\partial \mathbf{x}}(\tilde{\mathbf{x}}, \tilde{\lambda}) = 2\tilde{\mathbf{x}}^T Q - 2\tilde{\lambda}\tilde{\mathbf{x}}^T P$$

$$\frac{\partial L}{\partial \lambda}(\tilde{\mathbf{x}}, \tilde{\lambda}) = 1 - \tilde{\mathbf{x}}^T P \tilde{\mathbf{x}} .$$

Thus the necessary conditions are

$$Q\tilde{\mathbf{x}} = \tilde{\lambda} P \tilde{\mathbf{x}}$$
$$\tilde{\mathbf{x}}^T P \tilde{\mathbf{x}} = 1 .$$

Since P is nonsingular we obtain

$$P^{-1}Q\tilde{\mathbf{x}} = \tilde{\lambda}\tilde{\mathbf{x}}.$$

Since $\mathbf{x}^T P \mathbf{x} = 1$ the (column) vector $\mathbf{x}$ cannot be the zero vector. Thus the above equation is an eigenvalue equation, i.e. $\tilde{\lambda}$ is an eigenvalue of the $n \times n$ matrix $P^{-1}Q$. Multiplying this equation with the row vector $\tilde{\mathbf{x}}^T P$ we obtain the scalar equation

$$\tilde{\mathbf{x}}^T Q \tilde{\mathbf{x}} = \tilde{\lambda}\tilde{\mathbf{x}}^T P \tilde{\mathbf{x}} = \tilde{\lambda}.$$

Thus we conclude that the maximizer of $\mathbf{x}^T Q \mathbf{x}$ subject to the constraint $\mathbf{x}^T P \mathbf{x} = 1$ is the eigenvector of the matrix $P^{-1}Q$ corresponding to the largest eigenvalue.

Problem 8. Let A be an arbitrary $n \times n$ matrix over **C**. Let

$$H := \frac{A + A^*}{2}, \qquad S := \frac{A - A^*}{2i}.$$

Let λ be an eigenvalue of A and $\mathbf{x}$ be the corresponding normalized eigenvector (column vector).
(i) Show that

$$\lambda = \mathbf{x}^* H \mathbf{x} + i\mathbf{x}^* S \mathbf{x}.$$

(ii) Show that the real part λ_r of the eigenvalue λ is given by $\lambda_r = \mathbf{x}^* H \mathbf{x}$ and the imaginary part λ_i is given by $\lambda_i = \mathbf{x}^* S \mathbf{x}$.

Solution 8.   (i) The eigenvalue equation is given by $A\mathbf{x} = \lambda\mathbf{x}$. Then using $\mathbf{x}^* \mathbf{x} = 1$ we find

$$\begin{aligned}
\lambda &= \mathbf{x}^* A \mathbf{x} \\
&= \mathbf{x}^* \frac{A + A^*}{2}\mathbf{x} + \mathbf{x}^* \frac{A - A^*}{2}\mathbf{x} \\
&= \mathbf{x}^* H \mathbf{x} + i\mathbf{x}^* S \mathbf{x}.
\end{aligned}$$

(ii) Note that H and S are hermitian, i.e., $H^* = H$ and $S^* = S$. Thus we have

$$\lambda^* = \mathbf{x}^* H \mathbf{x} - i\mathbf{x}^* S \mathbf{x}.$$

Calculating $\lambda + \lambda^*$ and $\lambda - \lambda^*$ provides the results.

Problem 9. Let $A = (a_{jk})$ be a normal nonsymmetric 3×3 matrix over the real numbers. Show that

$$\mathbf{a} = \begin{pmatrix} a_1 \\ a_2 \\ a_3 \end{pmatrix} = \begin{pmatrix} a_{23} - a_{32} \\ a_{31} - a_{13} \\ a_{12} - a_{21} \end{pmatrix}$$

is an eigenvector of A.

Solution 9.   Since A is normal with the underlying field $\mathbf{R}$ we have $AA^T = A^T A$. Since A is nonsymmetric the matrix

$$S := A - A^T = \begin{pmatrix} 0 & a_3 & -a_2 \\ -a_3 & 0 & a_1 \\ a_2 & -a_1 & 0 \end{pmatrix}$$

is skew-symmetric with rank 2. The *kernel* of S is therefore one-dimensional. We have

$$A\mathbf{a} = \mathbf{0}.$$

Therefore $\mathbf{a}$ is an element of the kernel. Since A is normal, we find

$$AS = A(A - A^T) = A^2 - AA^T = A^2 - A^T A = (A - A^T)A = SA.$$

If we multiply the equation $\mathbf{0} = S\mathbf{a}$ with A we obtain

$$\mathbf{0} = A\mathbf{0} = A(S\mathbf{a}) = (AS)\mathbf{a} = (SA)\mathbf{a} = S(A\mathbf{a}).$$

Thus besides $\mathbf{a}$, $A\mathbf{a}$ also belongs to the one-dimensional kernel of S. Therefore $A\mathbf{a} = \lambda\mathbf{a}$.

Problem 10.   Let λ_1, λ_2 and λ_3 be the eigenvalues of the matrix

$$A = \begin{pmatrix} 0 & 1 & 2 \\ 0 & 0 & 1 \\ 2 & 2 & 1 \end{pmatrix}.$$

Find $\lambda_1^2 + \lambda_2^2 + \lambda_3^2$ without calculating the eigenvalues of A or A^2.

Solution 10.   From the eigenvalue equation $A\mathbf{x} = \lambda\mathbf{x}$ we obtain $A^2\mathbf{x} = \lambda^2\mathbf{x}$. Now

$$\text{tr}A = \lambda_1 + \lambda_2 + \lambda_3.$$

Thus

$$\text{tr}A^2 = \lambda_1^2 + \lambda_2^2 + \lambda_3^2.$$

Since the diagonal part of A^2 is given by (4 2 7) we find

$$\lambda_1^2 + \lambda_2^2 + \lambda_3^2 = 13.$$

Thus we only have to calculate the diagonal part of A^2.

Problem 11.   Find all solutions of the linear equation

$$\begin{pmatrix} \cos\theta & -\sin\theta \\ -\sin\theta & -\cos\theta \end{pmatrix} \mathbf{x} = \mathbf{x}, \qquad \theta \in \mathbf{R} \tag{1}$$

with the condition that $\mathbf{x} \in \mathbf{R}^2$ and $\mathbf{x}^T\mathbf{x} = 1$, i.e., the vector $\mathbf{x}$ must be normalized. What type of equation is (1)?

Solution 11.   We find

$$\mathbf{x} = \begin{pmatrix} \cos(\theta/2) \\ -\sin(\theta/2) \end{pmatrix}.$$

Obviously (1) is an eigenvalue equation and the eigenvalue is $+1$.

Problem 12.   Consider the column vectors $\mathbf{u}$ and $\mathbf{v}$ in $\mathbf{R}^n$

$$\mathbf{u} = \begin{pmatrix} \cos\theta \\ \cos(2\theta) \\ \vdots \\ \cos(n\theta) \end{pmatrix}, \qquad \mathbf{v} = \begin{pmatrix} \sin\theta \\ \sin(2\theta) \\ \vdots \\ \sin(n\theta) \end{pmatrix}$$

where $n \geq 3$ and $\theta = 2\pi/n$.
(i) Calculate $\mathbf{u}^T\mathbf{u} + \mathbf{v}^T\mathbf{v}$.
(ii) Calculate $\mathbf{u}^T\mathbf{u} - \mathbf{v}^T\mathbf{v} + 2i\mathbf{u}^T\mathbf{v}$.
(iii) Calculate the matrix $A = \mathbf{u}\mathbf{u}^T - \mathbf{v}\mathbf{v}^T$, $A\mathbf{u}$ and $A\mathbf{v}$. Give an interpretation of the results.

Solution 12.   (i) We obtain

$$\mathbf{u}^T\mathbf{u} + \mathbf{v}^T\mathbf{v} = \sum_{j=1}^{n}(\cos^2(j\theta) + \sin^2(j\theta)) = \sum_{j=1}^{n}1 = n.$$

(ii) We obtain

$$\mathbf{u}^T\mathbf{u} - \mathbf{v}^T\mathbf{v} + 2i\mathbf{u}^T\mathbf{v} = \sum_{j=1}^{n}(\cos^2(j\theta) - \sin^2(j\theta) + 2i\cos(j\theta)\sin(j\theta)) = 0.$$

Thus $\mathbf{u}^T\mathbf{u} = \mathbf{v}^T\mathbf{v}$ and and $\mathbf{u}^T\mathbf{v} = 0$. Using the result of (i) we find $\mathbf{u}^T\mathbf{u} = \mathbf{v}^T\mathbf{v} = n/2$.
(iii) Using the result from (ii) we find

$$A\mathbf{u} = (\mathbf{u}\mathbf{u}^T - \mathbf{v}\mathbf{v}^T)\mathbf{u} = \mathbf{u}(\mathbf{u}^T\mathbf{u}) - \mathbf{v}(\mathbf{v}^T\mathbf{u}) = \frac{n}{2}\mathbf{u}$$

$$A\mathbf{v} = (\mathbf{u}\mathbf{u}^T - \mathbf{v}\mathbf{v}^T)\mathbf{v} = \mathbf{u}(\mathbf{u}^T\mathbf{v}) - \mathbf{v}(\mathbf{v}^T\mathbf{v}) = -\frac{n}{2}\mathbf{v}.$$

Thus we conclude that A has the eigenvalues $n/2$ and $-n/2$.

Problem 13.   Let A be an $n \times n$ matrix over $\mathbf{C}$. We define

$$r_j := \sum_{\substack{k=1 \\ k\neq j}}^{n}|a_{jk}|, \quad j = 1, 2, \ldots, n.$$

(i) Show that each eigenvalue λ of A satisfies at least one of the following inequalities

$$|\lambda - a_{jj}| \le r_j, \quad j = 1, 2, \ldots, n.$$

In other words show that all eigenvalues of A can be found in the union of disks

$$\{\, z \,:\, |z - a_{jj}| \le r_j, \ j = 1, 2, \ldots, n \,\}$$

This is *Geršgorin disk theorem*.

(ii) Apply this theorem to the matrix

$$A = \begin{pmatrix} 0 & i \\ -i & 0 \end{pmatrix}.$$

(iii) Apply this theorem to the matrix

$$B = \begin{pmatrix} 1 & 2 & 3 \\ 3 & 4 & 9 \\ 1 & 1 & 1 \end{pmatrix}.$$

Solution 13. (i) From the eigenvalue equation $A\mathbf{x} = \lambda\mathbf{x}$ we obtain

$$(\lambda I_n - A)\mathbf{x} = \mathbf{0}.$$

Splitting the matrix A into its diagonal part and non-diagonal part we obtain

$$(\lambda - a_{jj})x_j = \sum_{\substack{k=1 \\ k \ne j}}^{n} a_{jk}x_k, \quad j = 1, 2, \ldots, n$$

where x_j is the jth component of the vector $\mathbf{x}$. Let x_k be the largest component (in absolute value) of the vector $\mathbf{x}$. Then, since $|x_j|/|x_k| \le 1$ for $j \ne k$, we obtain

$$|\lambda - a_{kk}| \le \sum_{\substack{j=1 \\ j \ne k}}^{n} |a_{kj}| \frac{|x_j|}{|x_k|} \le \sum_{\substack{j=1 \\ j \ne k}}^{n} |a_{kj}|.$$

Thus the eigenvalue λ is contained in the disk $\{\, \lambda \,:\, |\lambda - a_{kk}| \le r_k \,\}$.

(ii) Since $|i| = |-i| = 1$ we obtain $r_1 = 1$, $r_2 = 1$. Thus the Geršgorin disks are

$$R_1 : \ \{\, z \,:\, |z| \le 1 \,\}$$
$$R_2 : \ \{\, z \,:\, |z| \le 1 \,\}.$$

Since the matrix is hermitian the eigenvalues are real and therefore we have $|x| \le 1$ in both cases for x real. The eigenvalues of the matrix A are given by 1, -1.

(iii) We obtain $r_1 = 5$, $r_2 = 12$, $r_3 = 2$. The Geršgorin disks are

$$R_1 : \{ z : |z - 1| \le 5 \}$$
$$R_2 : \{ z : |z - 4| \le 12 \}$$
$$R_3 : \{ z : |z - 1| \le 2 \}.$$

The eigenvalues of the matrix B are

$$\lambda_1 = 7.3067, \quad \lambda_{2,3} = -0.6533 \pm 0.3473i.$$

Problem 14.   Let A be an $n \times n$ matrix over $\mathbf{C}$. Let f be an *entire function*, i.e., an analytic function on the whole complex plane, for example $\exp(z)$, $\sin(z)$, $\cos(z)$. An infinite series expansion for $f(A)$ is not generally useful for computing $f(A)$. Using the *Cayley-Hamilton theorem* we can write

$$f(A) = a_{n-1}A^{n-1} + a_{n-2}A^{n-2} + \cdots + a_2 A^2 + a_1 A + a_0 I_n \qquad (1)$$

where the complex numbers $a_0, a_1, \ldots, a_{n-1}$ are determined as follows:
Let

$$r(\lambda) := a_{n-1}\lambda^{n-1} + a_{n-2}\lambda^{n-2} + \cdots + a_2\lambda^2 + a_1\lambda + a_0$$

which is the right-hand side of (1) with A^j replaced by λ^j, where $j = 0, 1, \ldots, n-1$.
For each distinct eigenvalue λ_j of the matrix A, we consider the equation

$$f(\lambda_j) = r(\lambda_j). \qquad (2)$$

If λ_j is an eigenvalue of multiplicity k, for $k > 1$, then we consider also the following equations

$$f'(\lambda)|_{\lambda=\lambda_j} = r'(\lambda)|_{\lambda=\lambda_j}$$
$$f''(\lambda)|_{\lambda=\lambda_j} = r''(\lambda)|_{\lambda=\lambda_j}$$
$$\cdots = \cdots$$
$$f^{(k-1)}(\lambda)\Big|_{\lambda=\lambda_j} = r^{(k-1)}(\lambda)\Big|_{\lambda=\lambda_j}.$$

Apply this technique to find $\exp(A)$ with

$$A = \begin{pmatrix} c & c \\ c & c \end{pmatrix}, \quad c \in \mathbf{R}, \quad c \ne 0.$$

Solution 14. We have

$$e^A = a_1 A + a_0 I_2 = c \begin{pmatrix} a_1 & a_1 \\ a_1 & a_1 \end{pmatrix} + \begin{pmatrix} a_0 & 0 \\ 0 & a_0 \end{pmatrix} = \begin{pmatrix} a_0 + ca_1 & ca_1 \\ ca_1 & a_0 + ca_1 \end{pmatrix}.$$

The eigenvalues of A are 0 and $2c$. Thus we obtain the two linear equations

$$e^0 = 0a_1 + a_0 = a_0$$
$$e^{2c} = 2ca_1 + a_0.$$

Solving these two linear equations yields

$$a_0 = 1, \qquad a_1 = \frac{e^{2c} - 1}{2c}.$$

It follows that

$$e^A = \begin{pmatrix} a_0 + ca_1 & ca_1 \\ ca_1 & ca_1 + a_0 \end{pmatrix} = \begin{pmatrix} (e^{2c} + 1)/2 & (e^{2c} - 1)/2 \\ (e^{2c} - 1)/2 & (e^{2c} + 1)/2 \end{pmatrix}.$$

Problem 15. (i) Use the method given above to calculate $\exp(iK)$, where the hermitian 2×2 matrix K is given by

$$K = \begin{pmatrix} a & b \\ \bar{b} & c \end{pmatrix}, \qquad a, c \in \mathbf{R}, \quad b \in \mathbf{C}.$$

(ii) Find the condition on a, b and c such that

$$e^{iK} = \frac{1}{\sqrt{2}} \begin{pmatrix} 1 & 1 \\ 1 & -1 \end{pmatrix}.$$

Solution 15. (i) The characteristic equation for iK is given by

$$\lambda^2 + \lambda(-ia - ic) = ac - b\bar{b}.$$

Thus the eigenvalues are given by

$$\lambda_{1,2} = \frac{i(a + c)}{2} \pm \frac{1}{2} \sqrt{2ac - a^2 - c^2 - 4b\bar{b}}.$$

We set in the following

$$\Delta := \lambda_1 - \lambda_2 = \sqrt{2ac - a^2 - c^2 - 4b\bar{b}}.$$

To apply the method given above we have

$$r(\lambda) = \alpha_1 \lambda + \alpha_0 = f(\lambda) = e^{\lambda}.$$

Thus we obtain the two linear equations

$$e^{\lambda_1} = \alpha_1 \lambda_1 + \alpha_0, \qquad e^{\lambda_2} = \alpha_1 \lambda_2 + \alpha_0.$$

It follows that

$$\alpha_1 = \frac{e^{\lambda_1} - e^{\lambda_2}}{\lambda_1 - \lambda_2}, \qquad \alpha_0 = \frac{e^{\lambda_2} \lambda_1 - e^{\lambda_1} \lambda_2}{\lambda_1 - \lambda_2}.$$

Thus we have

$$e^{iK} = \alpha_1 iK + \alpha_0 I_2 = \begin{pmatrix} i\alpha_1 a + \alpha_0 & i\alpha_1 b \\ i\alpha_1 \bar{b} & i\alpha_1 c + \alpha_0 \end{pmatrix}.$$

(ii) From the condition

$$\begin{pmatrix} i\alpha_1 a + \alpha_0 & i\alpha_1 b \\ i\alpha_1 \bar{b} & i\alpha_1 c + \alpha_0 \end{pmatrix} = \frac{1}{\sqrt{2}} \begin{pmatrix} 1 & 1 \\ 1 & -1 \end{pmatrix}$$

we obtain the four equations

$$i\alpha_1 a + \alpha_0 = \frac{1}{\sqrt{2}}, \qquad i\alpha_1 c + \alpha_0 = -\frac{1}{\sqrt{2}}$$

$$i\alpha_1 b = \frac{1}{\sqrt{2}}, \qquad i\alpha_1 \bar{b} = \frac{1}{\sqrt{2}}.$$

From the last two equation we find that $\bar{b} = b$, i.e., b is real. From the first two equations we find

$$\alpha_0 = -\frac{i}{2}\alpha_1(a + c)$$

and therefore, using the last two equations, $c = a - 2b$. Thus

$$\begin{pmatrix} i\alpha_1 a + \alpha_0 & i\alpha_1 b \\ i\alpha_1 \bar{b} & i\alpha_1 c + \alpha_0 \end{pmatrix} = \begin{pmatrix} i\alpha_1 b & i\alpha_1 b \\ i\alpha_1 b & -i\alpha_1 b \end{pmatrix}.$$

From the eigenvalues of e^{iK} we find

$$e^{\lambda_1} - e^{\lambda_2} = 2$$

and

$$\Delta = \sqrt{2ac - a^2 - c^2 - 4b^2} = 2\sqrt{2}ib.$$

Furthermore

$$\lambda_1 = i(a - b) + \sqrt{2}ib, \qquad \lambda_2 = i(a - b) - \sqrt{2}ib.$$

Thus we arrive at the equation

$$e^{i(a-b)+\sqrt{2}ib} - e^{i(a-b)-\sqrt{2}ib} = 2.$$

It follows that

$$ie^{i(a-b)}\sin(\sqrt{2}b) = 1$$

and therefore

$$i\cos(a - b)\sin(\sqrt{2}b) - \sin(a - b)\sin(\sqrt{2}b) = 1$$

with a solution

$$b = \frac{\pi}{2\sqrt{2}}, \qquad a = \frac{\pi}{2}\left(3 + \frac{1}{\sqrt{2}}\right).$$

Thus

$$c = a - 2b = \frac{\pi}{2}\left(3 - \frac{1}{\sqrt{2}}\right).$$

Thus the matrix K is given by

$$K = \frac{\pi}{2}\begin{pmatrix} 3 + 1/\sqrt{2} & 1/\sqrt{2} \\ 1/\sqrt{2} & 3 - 1/\sqrt{2} \end{pmatrix} = \frac{3\pi}{2}\begin{pmatrix} 1 & 0 \\ 0 & 1 \end{pmatrix} + \frac{\pi}{2} \cdot \frac{1}{\sqrt{2}}\begin{pmatrix} 1 & 1 \\ 1 & -1 \end{pmatrix}.$$

Problem 16. Let A be a *normal* matrix over $\mathbf{C}$, i.e. $A^*A = AA^*$. Show that if $\mathbf{x}$ is an eigenvector of A with eigenvalue λ, then $\mathbf{x}$ is an eigenvector of A^* with eigenvalue $\bar{\lambda}$.

Solution 16. Since $AA^* = A^*A$ and $A^{**} = A$ we have

$$(A\mathbf{x})^*A\mathbf{x} = \mathbf{x}^*A^*A\mathbf{x} = \mathbf{x}^*AA^*\mathbf{x} = (A^*\mathbf{x})^*A^*\mathbf{x}.$$

It follows that

$$\begin{aligned}
0 = \mathbf{0}^*\mathbf{0} &= (A\mathbf{x} - \lambda\mathbf{x})^*(A\mathbf{x} - \lambda\mathbf{x}) \\
&= (\mathbf{x}^*A^* - \bar{\lambda}\mathbf{x}^*)(A\mathbf{x} - \lambda\mathbf{x}) \\
&= \mathbf{x}^*A^*A\mathbf{x} - \bar{\lambda}\mathbf{x}^*A\mathbf{x} - \lambda\mathbf{x}^*A^*\mathbf{x} + \lambda\bar{\lambda}\mathbf{x}^*\mathbf{x} \\
&= \mathbf{x}^*AA^*\mathbf{x} - \bar{\lambda}\mathbf{x}^*A\mathbf{x} - \lambda\mathbf{x}^*A^*\mathbf{x} + \lambda\bar{\lambda}\mathbf{x}^*\mathbf{x} \\
&= (A^*\mathbf{x} - \bar{\lambda}\mathbf{x})^*(A^*\mathbf{x} - \bar{\lambda}\mathbf{x}).
\end{aligned}$$

Thus, we have the eigenvalue equation $A^*\mathbf{x} = \bar{\lambda}\mathbf{x}$, which implies that $\mathbf{x}$ is an eigenvector of A^* corresponding to the eigenvalue $\bar{\lambda}$.

Problem 17. Show that an $n \times n$ matrix A is singular if and only if at least one eigenvalue is 0.

Solution 17. From the eigenvalue equation $A\mathbf{x} = \lambda\mathbf{x}$ we obtain

$$\det(A - \lambda I_n) = 0.$$

Thus if $\lambda = 0$ we obtain $\det A = 0$. It follows that A is singular, i.e. A^{-1} does not exist.

Problem 18. Let A be an invertible $n \times n$ matrix. Show that if $\mathbf{x}$ is an eigenvector of A with eigenvalue λ, then $\mathbf{x}$ is an eigenvector of A^{-1} with eigenvalue λ^{-1}.

Solution 18. From $A\mathbf{x} = \lambda\mathbf{x}$ we obtain

$$A^{-1}A\mathbf{x} = A^{-1}(\lambda\mathbf{x}).$$

Thus $\mathbf{x} = \lambda A^{-1}\mathbf{x}$. Since $\det A \neq 0$ all eigenvalues must be nonzero. Therefore

$$A^{-1}\mathbf{x} = \frac{1}{\lambda}\mathbf{x}.$$

Problem 19. Let A be an $n \times n$ matrix over $\mathbf{R}$. Show that A and A^T have the same eigenvalues.

Solution 19. Let λ be an eigenvalue of A. Then we have

$$0 = \det(A - \lambda I_n) = \det((A^T)^T - \lambda I_n^T) = \det(A^T - \lambda I_n)^T = \det(A^T - \lambda I_n)$$

since for any $n \times n$ matrix B we have $\det B = \det B^T$.

Problem 20. Let A be a symmetric matrix over $\mathbf{R}$. Since A is symmetric over $\mathbf{R}$ there exists a set of orthonormal eigenvectors $\mathbf{v}_1, \mathbf{v}_2, \ldots, \mathbf{v}_n$ which form an orthonormal basis. Let $\mathbf{x} \in \mathbf{R}^n$ be a reasonably good approximation to an eigenvector, say $\mathbf{v}_1$. Calculate

$$R := \frac{\mathbf{x}^T A\mathbf{x}}{\mathbf{x}^T\mathbf{x}}.$$

The quotient is called *Rayleigh quotient*. Discuss.

Solution 20. We can write

$$\mathbf{x} = c_1\mathbf{v}_1 + c_2\mathbf{v}_2 + \cdots + c_n\mathbf{v}_n, \qquad c_j \in \mathbf{R}.$$

Then, since $A\mathbf{v}_j = \lambda_j\mathbf{v}_j$, $j = 1, 2, \ldots, n$ and using $\mathbf{v}_j^T\mathbf{v}_k = 0$ if $j \neq k$ we find

$$
\begin{aligned}
R &= \frac{\mathbf{x}^T A\mathbf{x}}{\mathbf{x}^T\mathbf{x}} \\
&= \frac{(c_1\mathbf{v}_1 + \cdots + c_n\mathbf{v}_n)^T A(c_1\mathbf{v}_1 + \cdots + c_n\mathbf{v}_n)}{(c_1\mathbf{v}_1 + \cdots + c_n\mathbf{v}_n)^T(c_1\mathbf{v}_1 + \cdots + c_n\mathbf{v}_n)}
\end{aligned}
$$

$$= \frac{(c_1 \mathbf{v}_1 + \cdots + c_n \mathbf{v}_n)^T (c_1 \lambda_1 \mathbf{v}_1 + \cdots + c_n \lambda_n \mathbf{v}_n)}{c_1^2 + c_2^2 + \cdots + c_n^2}$$

$$= \frac{\lambda_1 c_1^2 + \lambda_2 c_2^2 + \cdots + \lambda_n c_n^2}{c_1^2 + c_2^2 + \cdots + c_n^2}$$

$$= \lambda_1 \left(\frac{1 + (\lambda_2/\lambda_1)(c_2/c_1)^2 + \cdots + (\lambda_n/\lambda_1)(c_n/c_1)^2}{1 + (c_2/c_1)^2 + \cdots + (c_n/c_1)^2} \right).$$

Since $\mathbf{x}$ is a good approximation to $\mathbf{v}_1$, the coefficient c_1 is larger than the other c_j, $j = 2, 3, \ldots, n$. Thus the expression in the parenthesis is close to 1, which means that R_q is close to λ_1.

Problem 21. Given a 4×4 symmetric matrix A over $\mathbf{R}$ with the eigenvalues $\lambda_1 = 0$, $\lambda_2 = 1$, $\lambda_3 = 2$, $\lambda_4 = 3$ and the corresponding normalized eigenvectors

$$\mathbf{x}_1 = \frac{1}{\sqrt{2}} \begin{pmatrix} 1 \\ 0 \\ 0 \\ 1 \end{pmatrix}, \quad \mathbf{x}_2 = \frac{1}{\sqrt{2}} \begin{pmatrix} 1 \\ 0 \\ 0 \\ -1 \end{pmatrix}, \quad \mathbf{x}_3 = \frac{1}{\sqrt{2}} \begin{pmatrix} 0 \\ 1 \\ 1 \\ 0 \end{pmatrix}, \quad \mathbf{x}_4 = \frac{1}{\sqrt{2}} \begin{pmatrix} 0 \\ 1 \\ -1 \\ 0 \end{pmatrix}.$$

Reconstruct the matrix A using the spectral theorem. The eigenvectors given above are called *Bell basis*.

Solution 21. Using the *spectral theorem* we obtain

$$A = \sum_{j=1}^{4} \lambda_j \mathbf{x}_j \mathbf{x}_j^T = \mathbf{x}_2 \mathbf{x}_2^T + 2 \mathbf{x}_3 \mathbf{x}_3^T + 3 \mathbf{x}_4 \mathbf{x}_4^T.$$

Straigthforward calculation yields

$$A = \begin{pmatrix} 1/2 & 0 & 0 & -1/2 \\ 0 & 5/2 & -1/2 & 0 \\ 0 & -1/2 & 5/2 & 0 \\ -1/2 & 0 & 0 & 1/2 \end{pmatrix}.$$

Problem 22. Let A be an $n \times n$ hermitian matrix. Then $(A + iI_n)^{-1}$ exists and

$$U = (A - iI_n)(A + iI_n)^{-1} \tag{1}$$

is a unitary matrix (so-called *Cayley transform* of A).
(i) Show that $+1$ cannot be an eigenvalue of U.
(ii) Show that

$$A = i(U + I_n)(I_n - U)^{-1}.$$

Solution 22.   (i) Assume that $+1$ is an eigenvalue, i.e., $U\mathbf{x} = \mathbf{x}$, where $\mathbf{x}$ is the eigenvector. Then

$$(A - iI_n)(A + iI_n)^{-1}\mathbf{x} = \mathbf{x} = I_n\mathbf{x} = (A + iI_n)(A + iI_n)^{-1}\mathbf{x}.$$

Thus

$$2iI_n(A + iI_n)^{-1}\mathbf{x} = 0$$

or $(A + iI_n)^{-1}\mathbf{x} = 0$. Therefore $\mathbf{x} = 0$ (contradiction).
(ii) We set $\mathbf{y} = (A + iI_n)^{-1}\mathbf{x}$, where $\mathbf{x} \in \mathbf{C}^n$ is arbitrary. Then $\mathbf{x} = (A + iI_n)\mathbf{y}$. From (1) we obtain

$$U\mathbf{x} = (A - iI_n)(A + iI_n)^{-1}\mathbf{x}$$

and therefore

$$U\mathbf{x} = (A - iI_n)\mathbf{y}.$$

Adding and subtracting the equation $\mathbf{x} = (A + iI_n)\mathbf{y}$ yields the two equations

$$(U + I_n)\mathbf{x} = 2A\mathbf{y}, \qquad (I_n - U)\mathbf{x} = 2iI_n\mathbf{y}.$$

Thus

$$(U + I_n)\mathbf{x} = -iA(I_n - U)\mathbf{x}.$$

It follows that

$$(U + I_n) = -iA(I_n - U)$$
$$(U + I_n)(I_n - U)^{-1} = -iA.$$

Therefore

$$A = i(U + I_n)(I_n - U)^{-1}.$$

Problem 23.   Let A be an $n \times n$ real symmetric matrix and

$$Q(\mathbf{x}) := \mathbf{x}^T A \mathbf{x}.$$

The following statements hold (*maximum principle*)
1) $\lambda_1 = \max_{\|\mathbf{x}\|=1} Q(\mathbf{x}) = Q(\mathbf{x}_1)$ is the largest eigenvalue of the matrix A and $\mathbf{x}_1$ is the eigenvector corresponding to eigenvalue λ_1.
2) (inductive statement). Let $\lambda_k = \max Q(\mathbf{x})$ subject to the constraints
a) $\mathbf{x}^T\mathbf{x}_j = 0$, $j = 1, 2, \ldots, k - 1$.
b) $\|\mathbf{x}\| = 1$.
c) Then $\lambda_k = Q(\mathbf{x}_k)$ is the kth eigenvalue of A, $\lambda_1 \geq \lambda_2 \geq \ldots \geq \lambda_k$ and $\mathbf{x}_k$ is the corresponding eigenvectors of A.
Apply the maximum principle to the matrix

$$A = \begin{pmatrix} 3 & 1 \\ 1 & 3 \end{pmatrix}.$$

Solution 23. We find

$$Q(\mathbf{x}) = 3x_1^2 + 2x_1 x_2 + 3x_2^2.$$

We change to *polar coordinates* by setting $x_1 = \cos\theta$, $x_2 = \sin\theta$. Thus

$$Q(\theta) = 3 + 2\sin(2\theta).$$

Clearly $\max Q(\theta) = 4$ which occurs at $\theta = \pi/4$ and $\min Q(\theta) = 2$ at $\theta = -\pi/4$. It follows that $\lambda_1 = 4$ with $\mathbf{x}_1 = (1, 1)^T$ and $\lambda_2 = 2$ with $\mathbf{x}_2 = (1, -1)^T$.

Problem 24. Let A be an $n \times n$ matrix. An $n \times n$ matrix can have at most n linearly independent eigenvectors. Now assume that A has $n + 1$ eigenvectors (at least one must be linearly dependent) such that any n of them are linearly independent. Show that A is a scalar multiple of the identity matrix I_n.

Solution 24. Let $\mathbf{x}_1, \ldots, \mathbf{x}_{n+1}$ be the eigenvectors of A with eigenvalues $\lambda_1, \ldots, \lambda_{n+1}$. Since $\mathbf{x}_1, \ldots, \mathbf{x}_n$ are linearly independent, they span the vector space. Consequently

$$\mathbf{x}_{n+1} = \sum_{i=1}^n \alpha_i \mathbf{x}_i. \tag{1}$$

Multiplying equation (1) with λ_{n+1} we have

$$\lambda_{n+1} \mathbf{x}_{n+1} = \sum_{i=1}^n \alpha_i \lambda_{n+1} \mathbf{x}_i.$$

Applying A to equation (1) yields

$$\lambda_{n+1} \mathbf{x}_{n+1} = \sum_{i=1}^n \alpha_i \lambda_i \mathbf{x}_i.$$

Thus

$$\sum_{i=1}^n \alpha_i \lambda_{n+1} \mathbf{x}_i = \sum_{i=1}^n \alpha_i \lambda_i \mathbf{x}_i.$$

It follows that $\alpha_i \lambda_{n+1} = \alpha_i \lambda_i$ for all $i = 1, 2, \ldots, n$. If $\alpha_i = 0$ for some i, then $\mathbf{x}_{n+1}$ can be expressed as a linear combination of $\mathbf{x}_1, \ldots, \mathbf{x}_{i-1}, \mathbf{x}_{i+1}, \ldots, \mathbf{x}_n$, contradicting the linear independence. Therefore $\alpha_i \neq 0$ for all i, so that $\lambda_{n+1} = \lambda_i$ for all i. This implies $A = \lambda_{n+1} I_n$.

Problem 25. An $n \times n$ *stochastic matrix* P satisfies the following conditions:

$$p_{ij} \geq 0 \quad \text{for all } i, j = 1, 2, \ldots, n$$

and

$$\sum_{i=1}^{n} p_{ij} = 1 \quad \text{for all } j = 1, 2, \ldots, n.$$

Show that a stochastic matrix always has at least one eigenvalue equal to one.

Solution 25. Assume that 1 is an eigenvalue. Then the eigenvalue equation is given by

$$\begin{pmatrix} p_{11} & p_{12} & p_{13} & \cdots & p_{1n} \\ p_{21} & p_{22} & p_{23} & \cdots & p_{2n} \\ \vdots & \vdots & \vdots & \ddots & \vdots \\ p_{n1} & p_{n2} & p_{n3} & \cdots & p_{nn} \end{pmatrix} \begin{pmatrix} x_1 \\ x_2 \\ \vdots \\ x_n \end{pmatrix} = \begin{pmatrix} x_1 \\ x_2 \\ \vdots \\ x_n \end{pmatrix}.$$

From the second condition given above we find that the eigenvector **x** is

$$\mathbf{x} = \begin{pmatrix} 1 \\ 1 \\ \vdots \\ 1 \end{pmatrix}.$$

Thus 1 is an eigenvalue.

Problem 26. Let A be an $n \times n$ matrix over **C**. The *spectral radius* $\rho(A)$ is the radius of the smallest circle in the complex plane that contains all its eigenvalues. Every characteristic polynomial has at least one root. For any two $n \times n$ matrices $A = (a_{ij})$ and $B = (b_{ij})$, the *Hadamard product* of A and B is the $n \times n$ matrix

$$A \bullet B := (a_{ij}b_{ij}).$$

Let A, B be nonnegative matrices. Then

$$\rho(A \bullet B) \leq \rho(A)\rho(B).$$

Apply this inequality to the nonnegative matrices

$$A = \begin{pmatrix} 1/4 & 0 \\ 0 & 3/4 \end{pmatrix}, \qquad B = \begin{pmatrix} 1 & 1 \\ 1 & 1 \end{pmatrix}.$$

Solution 26. The eigenvalues of A are $1/4$ and $3/4$. Thus the spectral radius is $3/4$. The eigenvalues of B are 0 and 2. Thus the spectral radius is 2. Now

$$A \bullet B = \begin{pmatrix} 1/4 & 0 \\ 0 & 3/4 \end{pmatrix}.$$

The eigenvalues of $A \bullet B$ are $1/4$ and $3/4$ with the spectral radius is $3/4$. Thus $\rho(A \bullet B) = 3/4 \leq 3/4 \cdot 2 = 3/2$.

Problem 27. Let A be an $n \times n$ matrix over $\mathbf{C}$. Assume that A is hermitian and unitary. What can be said about the eigenvalues of A?

Solution 27. Since A is hermitian the eigenvalues are real. Since A is unitary the eigenvalues lie on the unit circle in the complex plane. Now A is hermitian and unitary we have that $\lambda \in \{+1, -1\}$.

Problem 28. Consider the $(n + 1) \times (n + 1)$ matrix

$$A = \begin{pmatrix} 0 & \mathbf{s}^* \\ \mathbf{r} & 0_{n \times n} \end{pmatrix}$$

where $\mathbf{r}$ and $\mathbf{s}$ are $n \times 1$ vectors with complex entries, $\mathbf{s}^*$ denoting the conjugate transpose of $\mathbf{s}$. Find $\det(B - \lambda I_{n+1})$, i.e. find the characteristic polynomial.

Solution 28. Straightforward calculation yields

$$\det(B - \lambda I_{n+1}) = (-\lambda)^{n-1}(\lambda^2 - \mathbf{s}^*\mathbf{r}).$$

Problem 29. The matrix difference equation

$$\mathbf{p}(t + 1) = M\mathbf{p}(t), \qquad t = 0, 1, 2, \ldots$$

with the column vector (vector of probabilities)

$$\mathbf{p}(t) = (p_1(t), p_2(t), \ldots, p_n(t))^T$$

and the $n \times n$ matrix

$$M = \begin{pmatrix} (1 - w) & 0.5w & 0 & \cdots & 0.5w \\ 0.5w & (1 - w) & 0.5w & \cdots & 0 \\ 0 & 0.5w & (1 - w) & \cdots & 0 \\ \vdots & \vdots & \vdots & \ddots & \vdots \\ 0.5w & 0 & 0 & \cdots & (1 - w) \end{pmatrix}$$

plays a role in random walk in one dimension. M is called the *transition matrix* and w denotes the probability $w \in [0, 1]$ that at a given time step the particle jumps to either of its nearest neighbor sites, then the probability that the particle does not jump either to the right of left is $(1 - w)$. The matrix M is of the type known as *circulant matrix*. Such an $n \times n$ matrix is of the form

$$C = \begin{pmatrix} c_0 & c_1 & c_2 & \cdots & c_{n-1} \\ c_{n-1} & c_0 & c_1 & \cdots & c_{n-2} \\ c_{n-2} & c_{n-1} & c_0 & \cdots & c_{n-3} \\ \vdots & \vdots & \vdots & \ddots & \vdots \\ c_1 & c_2 & c_3 & \cdots & c_0 \end{pmatrix}$$

with the normalized eigenvectors

$$e_j = \frac{1}{\sqrt{n}} \begin{pmatrix} 1 \\ e^{2\pi ij/n} \\ \vdots \\ e^{2(n-1)\pi ij/n} \end{pmatrix}$$

for $j = 1, 2, \ldots, n$.

(i) Use this result to find the eigenvalues of the matrix C.

(ii) Use (i) to find the eigenvalues of the matrix M.

(iii) Use (ii) to find $\mathbf{p}(t)$ $(t = 0, 1, 2, \ldots)$, where we expand the initial distribution vector $\mathbf{p}(0)$ in terms of the eigenvectors

$$\mathbf{p}(0) = \sum_{k=1}^{n} a_k e_k$$

with

$$\sum_{j=1}^{n} p_j(0) = 1.$$

(iv) Assume that

$$\mathbf{p}(0) = \frac{1}{n} \begin{pmatrix} 1 \\ 1 \\ \vdots \\ 1 \end{pmatrix}.$$

Give the time evolution of $\mathbf{p}(0)$.

Solution 29. (i) Calculating Ce_j provides the eigenvalues

$$\lambda_j = c_0 + c_1 \exp(2\pi ij/n) + c_2 \exp(4\pi ij/n) + \cdots + c_{n-1} \exp(2(n-1)\pi ij/n).$$

(ii) We have $c_0 = 1 - w$, $c_1 = 0.5w$, $c_{n-1} = 0.5w$ and

$$c_2 = c_3 = \cdots = c_{n-2} = 0.$$

Using $e^{2\pi ij} = 1$ and the identity

$$\cos \alpha \equiv \frac{1}{2}(e^{i\alpha} + e^{-i\alpha})$$

we obtain as eigenvalues of M

$$\lambda_j = (1 - w) + w \cos(2j\pi/n)$$

for $j = 1, 2, \ldots, n$.

(iii) Since $\mathbf{p}(0)$ is a probability vector we have

$$p_j(0) \geq 0$$

and $\sum_{j=1}^{n} p_j(0) = 1$. The expansion coefficients a_k are given by

$$a_k = \frac{1}{\sqrt{n}} \sum_{j=1}^{n} p_j(0) \exp(-2\pi i(j - 1)k/n).$$

Thus we obtain

$$\mathbf{p}(t) = \sum_{k=1}^{n} a_k \lambda_k^t \mathbf{e}_k \qquad t = 0, 1, 2, \ldots .$$

(iv) Since

$$M\mathbf{p}(0) - \mathbf{p}(0)$$

we find that the probability vector $\mathbf{p}(0)$ is a *fixed point*.

Problem 30. Let U be a unitary matrix and $\mathbf{x}$ an eigenvector of U with the corresponding eigenvalue λ, i.e.

$$U\mathbf{x} = \lambda\mathbf{x}.$$

(i) Show that $U^*\mathbf{x} = \overline{\lambda}\mathbf{x}$.

(ii) Let λ, μ be distinct eigenvalues of a unitary matrix U with the corresponding eigenvectors $\mathbf{x}$ and $\mathbf{y}$, i.e.

$$U\mathbf{x} = \lambda\mathbf{x}, \qquad U\mathbf{y} = \mu\mathbf{y}.$$

Show that $\mathbf{x}^*\mathbf{y} = 0$.

Solution 30. (i) From $U\mathbf{x} = \lambda\mathbf{x}$ we obtain

$$U^*U\mathbf{x} = \lambda U^*\mathbf{x}.$$

Therefore $\mathbf{x} = \lambda U^*\mathbf{x}$. Finally $U^*\mathbf{x} = \overline{\lambda}\mathbf{x}$ since $\overline{\lambda}\lambda = 1$. Now λ can be written as $e^{i\alpha}$ with $\alpha \in \mathbf{R}$ we have $\overline{\lambda} = e^{-i\alpha}$.

(ii) Using the result from (i) we have

$$\mu\mathbf{x}^*\mathbf{y} = \mathbf{x}^*(\mu\mathbf{y}) = \mathbf{x}^*U\mathbf{y} = (U^*\mathbf{x})^*\mathbf{y} = (\overline{\lambda}\mathbf{x})^*\mathbf{y} = \lambda\mathbf{x}^*\mathbf{y}.$$

Thus

$$(\mu - \lambda)\mathbf{x}^*\mathbf{y} = 0.$$

Since μ and λ are distinct it follows that $\mathbf{x}^*\mathbf{y} = 0$.

Problem 31. Let H, H_0, V be $n \times n$ matrices over $\mathbf{C}$ and $H = H_0 + V$. Let $z \in \mathbf{C}$ and assume that z is chosen so that $(H_0 - zI_n)^{-1}$ and $(H - zI_n)^{-1}$ exist. Show that

$$(H - zI_n)^{-1} = (H_0 - zI_n)^{-1} - (H_0 - zI_n)^{-1}V(H - zI_n)^{-1}.$$

This is called the second *resolvent identity*.

Solution 31.   We have

$$H_0 + V = H$$
$$\Leftrightarrow H_0 = H - V$$
$$\Leftrightarrow (H_0 - zI_n) = (H - zI_n) - V$$
$$\Leftrightarrow (H_0 - zI_n)^{-1}(H_0 - zI_n) = (H_0 - zI_n)^{-1}(H - zI_n) - (H_0 - zI_n)^{-1}V$$
$$\Leftrightarrow I_n = (H_0 - zI_n)^{-1}(H - zI_n) - (H_0 - zI_n)^{-1}V$$
$$(H - zI_n)^{-1} = (H_0 - zI_n)^{-1} - (H_0 - zI_n)^{-1}V(H - zI_n)^{-1}.$$

Problem 32.   Let $\mathbf{u}$ be a nonzero column vector in $\mathbf{R}^n$. Consider the $n \times n$ matrix

$$A = \mathbf{u}\mathbf{u}^T - \mathbf{u}^T\mathbf{u}I_n.$$

Is $\mathbf{u}$ an eigenvector of this matrix? If so what is the eigenvalue?

Solution 32.   We have

$$(\mathbf{u}\mathbf{u}^T - \mathbf{u}^T\mathbf{u}I_n)\mathbf{u} = \mathbf{u}(\mathbf{u}^T\mathbf{u}) - (\mathbf{u}^T\mathbf{u})\mathbf{u} = (\mathbf{u}^T\mathbf{u})(\mathbf{u} - \mathbf{u}) = \mathbf{0}.$$

Thus $\mathbf{u}$ is an eigenvector with eigenvalue 0.

Problem 33.   An $n \times n$ matrix A is called a *Hadamard matrix* if each entry of A is 1 or -1 and if the rows or columns of A are orthogonal, i.e.,

$$AA^T = nI_n \quad \text{or} \quad A^TA = nI_n.$$

Note that $AA^T = nI_n$ and $A^TA = nI_n$ are equivalent. Hadamard matrices H_n of order 2^n can be generated recursively by defining

$$H_1 = \begin{pmatrix} 1 & 1 \\ 1 & -1 \end{pmatrix}, \qquad H_n = \begin{pmatrix} H_{n-1} & H_{n-1} \\ H_{n-1} & -H_{n-1} \end{pmatrix}$$

for $n \geq 2$. Show that the eigenvalues of H_n are given by $+2^{n/2}$ and $-2^{n/2}$ each of multiplicity 2^{n-1}.

Solution 33. We use induction on n. The case $n = 1$ is obvious. Now for $n \geq 2$ we have

$$\det(\lambda I - H_n) = \begin{vmatrix} \lambda I - H_{n-1} & -H_{n-1} \\ -H_{n-1} & \lambda I + H_{n-1} \end{vmatrix}$$
$$= \det((\lambda I - H_{n-1})(\lambda I + H_{n-1}) - H_{n-1}^2).$$

Thus

$$\det(\lambda I - H_n) = \det(\lambda^2 I - 2H_{n-1}^2)$$
$$= \det(\lambda I - \sqrt{2}H_{n-1}) \det(\lambda I + \sqrt{2}H_{n-1}).$$

This shows that each eigenvalue μ of H_{n-1} generates two eigenvalues $\pm\sqrt{2\mu}$ of H_n. The assertion then follows by the induction hypothesis, for H_{n-1} has eigenvalues $+2^{(n-1)/2}$ and $-2^{(n-1)/2}$ each of multiplicity 2^{n-2}.

Problem 34. Let U be an $n \times n$ unitary matrix. Then U can be written as

$$U = V \operatorname{diag}(\lambda_1, \lambda_2, \dots, \lambda_n)V^*$$

where $\lambda_1, \lambda_2, \dots, \lambda_n$ are the eigenvalues of U and V is an $n \times n$ unitary matrix. Let

$$U = \begin{pmatrix} 0 & 1 \\ 1 & 0 \end{pmatrix}.$$

Find the decomposition for U given above.

Solution 34. The eigenvalues of U are $+1$ and -1. Thus we have

$$U = V \operatorname{diag}(1, -1)V^*$$

with

$$V = \frac{1}{\sqrt{2}} \begin{pmatrix} 1 & 1 \\ 1 & -1 \end{pmatrix}.$$

Therefore $V = V^*$.

Problem 35. An $n \times n$ matrix A over the complex numbers is called *positive semidefinite* (written as $A \geq 0$), if

$$\mathbf{x}^* A \mathbf{x} \geq 0 \quad \text{for all } \mathbf{x} \in \mathbf{C}^n.$$

Show that for every $A \geq 0$, there exists a unique $B \geq 0$ so that $B^2 = A$.

Solution 35. Let
$$A = U^* \text{diag}(\lambda_1, \ldots, \lambda_n)U$$
where U is unitary. We take
$$B = U^* \text{diag}(\lambda_1^{1/2}, \ldots, \lambda_n^{1/2})U.$$

Then the matrix B is positive semidefinite and $B^2 = A$ since $U^*U = I_n$. To show the uniqueness, suppose that C is an $n \times n$ positive semidefinite matrix satisfying $C^2 = A$. Since the eigenvalues of C are the nonnegative square roots of the eigenvalues of A, we can write
$$C = V \text{diag}(\lambda_1^{1/2}, \ldots, \lambda_n^{1/2})V^*$$
for some unitary matrix V. Then the identity
$$C^2 = A = B^2$$
yields
$$T \text{diag}(\lambda_1, \ldots, \lambda_n) = \text{diag}(\lambda_1, \ldots, \lambda_n)T$$
where $T = UV$. This yields
$$t_{jk}\lambda_k = \lambda_j t_{jk}.$$

Thus
$$t_{jk}\lambda_k^{1/2} = \lambda_j^{1/2}t_{jk}.$$

Hence
$$T \text{diag}(\lambda_1^{1/2}, \ldots, \lambda_n^{1/2}) = \text{diag}(\lambda_1^{1/2}, \ldots, \lambda_n^{1/2})T.$$
Since $T = UV$ it follows that $B = C$.

Problem 36. An $n \times n$ matrix A over the complex numbers is said to be *normal* if it commutes with its conjugate transpose $A^*A = AA^*$. The matrix A can be written
$$A = \sum_{j=1}^{n} \lambda_j E_j$$
where $\lambda_j \in \mathbb{C}$ are the eigenvalues of A and E_j are $n \times n$ matrices satisfying

$$E_j^2 = E_j = E_j^*, \qquad E_j E_k = 0_n \text{ if } j \neq k, \qquad \sum_{j=1}^{n} E_j = I_n.$$

Let
$$A = \begin{pmatrix} 0 & 1 \\ 1 & 0 \end{pmatrix}.$$

Find the decomposition of A given above.

Solution 36.   The eigenvalues of A are given by

$$\lambda_1 = +1, \qquad \lambda_2 = -1.$$

The matrices E_j are constructed from the normalized eigenvectors of A. The normalized eigenvectors of A are given by

$$\mathbf{x}_1 = \frac{1}{\sqrt{2}} \begin{pmatrix} 1 \\ 1 \end{pmatrix}, \qquad \mathbf{x}_2 = \frac{1}{\sqrt{2}} \begin{pmatrix} 1 \\ -1 \end{pmatrix}.$$

Thus

$$E_1 = \mathbf{x}_1 \mathbf{x}_1^* = \frac{1}{2} \begin{pmatrix} 1 & 1 \\ 1 & 1 \end{pmatrix}, \qquad E_2 = \mathbf{x}_2 \mathbf{x}_2^* = \frac{1}{2} \begin{pmatrix} 1 & -1 \\ -1 & 1 \end{pmatrix}$$

with

$$E_1 E_2 = \begin{pmatrix} 0 & 0 \\ 0 & 0 \end{pmatrix}.$$

Problem 37.   Let A be an $n \times n$ matrix over $\mathbf{R}$. Assume that A^{-1} exists. Let $\mathbf{u}, \mathbf{v} \in \mathbf{R}^n$, where $\mathbf{u}, \mathbf{v}$ are considered as column vectors.
(i) Show that if

$$\mathbf{v}^T A^{-1} \mathbf{u} = -1$$

then $A + \mathbf{u}\mathbf{v}^T$ is not invertible.
(ii) Assume that $\mathbf{v}^T A^{-1} \mathbf{u} \neq -1$. Show that

$$(A + \mathbf{u}\mathbf{v}^T)^{-1} = A^{-1} - \frac{A^{-1}\mathbf{u}\mathbf{v}^T A^{-1}}{1 + \mathbf{v}^T A^{-1}\mathbf{u}}.$$

Solution 37.   (i) From $\mathbf{v}^T A^{-1} \mathbf{u} = -1$ it follows that $\mathbf{u} \neq 0$, $\mathbf{v} \neq 0$ and $A^{-1}\mathbf{u} \neq 0$. Now we have

$$(A + \mathbf{u}\mathbf{v}^T)A^{-1}\mathbf{u} = AA^{-1}\mathbf{u} + \mathbf{u}\mathbf{v}^T A^{-1}\mathbf{u}$$
$$= \mathbf{u} + \mathbf{u}(\mathbf{v}^T A^{-1}\mathbf{u})$$
$$= \mathbf{u} - \mathbf{u}$$
$$= 0$$

where we used $\mathbf{v}^T A^{-1}\mathbf{u} = -1$. Now

$$(A + \mathbf{u}\mathbf{v}^T)(A^{-1}\mathbf{u}) = 0$$

is an eigenvalue equation with eigenvalue 0 and eigenvector $A^{-1}\mathbf{u}$. Since

$$\det(A + \mathbf{uv}^T) = \lambda_1 \cdot \lambda_2 \cdot \ldots \cdot \lambda_n$$

where $\lambda_1, \lambda_2, \ldots, \lambda_n$ are the eigenvalues of $A + \mathbf{uv}^T$, we have $\det(A + \mathbf{uv}^T) = 0$. Thus the matrix $A + \mathbf{uv}^T$ is not invertible.

(ii) Since $\mathbf{v}^T A^{-1} \mathbf{u} \in \mathbf{R}$ we have the identity

$$I_n = I_n + \mathbf{uv}^T A^{-1} (\mathbf{v}^T A^{-1} \mathbf{u}) - \mathbf{u}(\mathbf{v}^T A^{-1} \mathbf{u})\mathbf{v}^T A^{-1}$$

where I_n is the $n \times n$ identity matrix. Thus

$$I_n = I_n + \mathbf{uv}^T A^{-1} - \mathbf{uv}^T A^{-1} + \mathbf{uv}^T A^{-1} \mathbf{v}^T A^{-1} \mathbf{u} - \mathbf{uv}^T A^{-1} \mathbf{uv}^T A^{-1}$$

and therefore

$$(A + \mathbf{uv}^T)(A + \mathbf{uv}^T)^{-1} = (A + \mathbf{uv}^T)A^{-1} - \frac{(A + \mathbf{uv}^T)A^{-1}\mathbf{uv}^T A^{-1}}{1 + \mathbf{v}^T A^{-1}\mathbf{u}}.$$

Finally

$$(A + \mathbf{uv}^T)^{-1} = A^{-1} - \frac{A^{-1}\mathbf{uv}^T A^{-1}}{1 + \mathbf{v}^T A^{-1}\mathbf{u}}.$$

Problem 38.   The *Denman-Beavers iteration* for the square root of an $n \times n$ matrix A with no eigenvalues on $\mathbf{R}^-$ is

$$Y_{k+1} = \frac{1}{2}(Y_k + Z_k^{-1})$$

$$Z_{k+1} = \frac{1}{2}(Z_k + Y_k^{-1})$$

with $k = 0, 1, 2, \ldots$ and $Z_0 = I_n$ and $Y_0 = A$. The iteration has the properties that

$$\lim_{k \to \infty} Y_k = A^{1/2}, \qquad \lim_{k \to \infty} Z_k = A^{-1/2}$$

and, for all k,

$$Y_k = AZ_k, \quad Y_k Z_k = Z_k Y_k, \quad Y_{k+1} = \frac{1}{2}(Y_k + AY_k^{-1}).$$

(i) Can the Denman-Beavers iteration be applied to the matrix

$$A = \begin{pmatrix} 1 & 1 \\ 1 & 2 \end{pmatrix}?$$

(ii) Find Y_1 and Z_1.

Solution 38. (i) For the eigenvalues we find

$$\lambda_{1,2} = \frac{1}{2}(3 \pm \sqrt{5}).$$

Thus there is no eigenvalue on $\mathbf{R}^-$ and the Denman-Beavers iteration can be applied.

(ii) We have $\det A = 1$. For the inverse of A we find

$$A^{-1} = \begin{pmatrix} 2 & -1 \\ -1 & 1 \end{pmatrix}$$

and therefore

$$Y_1 = \begin{pmatrix} 1 & 1/2 \\ 1/2 & 3/2 \end{pmatrix}, \qquad Z_1 = \begin{pmatrix} 3/2 & -1/2 \\ -1/2 & 1/2 \end{pmatrix}.$$

For $n \to \infty$, Y_n converges to

$$\begin{pmatrix} 0.89443 & 0.44721 \\ 0.44721 & 1.34164 \end{pmatrix}.$$

Problem 39. Let

$$A = \begin{pmatrix} 3 & 2 \\ 4 & 3 \end{pmatrix}$$

and I_2 be the 2×2 identity matrix. For $j \geq 1$, let d_j be the greatest common divisor of the entries of $A^j - I_2$. Show that

$$\lim_{j \to \infty} d_j = \infty.$$

Hint. Use the eigenvalues of A and the characteristic polynomial.

Solution 39. We have $\det A = 1$ and thus $1 = \lambda_1 \lambda_2$, where λ_1 and λ_2 denote the eigenvalues of A. The characteristic polynomial of A is given by $x^2 - 6x + 1$ with the eigenvalues

$$\lambda_1 = 3 + 2\sqrt{2}, \qquad \lambda_2 = \frac{1}{\lambda_1} = 3 - 2\sqrt{2}.$$

Therefore there exists an invertible matrix C such that $A = CDC^{-1}$ with

$$D = \begin{pmatrix} \lambda_1 & 0 \\ 0 & 1/\lambda_1 \end{pmatrix}$$

and the entries of the matrix C are in $\mathbf{Q}(\sqrt{2})$. We choose an integer $k \geq 1$ such that the entries of kC and kC^{-1} are in $\mathbf{Z}(\sqrt{2})$. Then

$$k^2(A^j - I_2) = (kC)(D^j - I_2)(kC^{-1})$$

and

$$D^j - I_2 = (\lambda_1^j - 1) \begin{pmatrix} 1 & 0 \\ 0 & \lambda_1^{-j} \end{pmatrix}.$$

Thus $\lambda_1^j - 1$ divides $k^2 d_j$ in $\mathbf{Z}(\sqrt{2})$. Taking norms, we find that the integer $(\lambda_1^j - 1)(\lambda_1^{-j} - 1)$ divides $k^4 d_j^2$. However $|\lambda_1| > 1$, so

$$|(\lambda_1^j - 1)(\lambda_1^{-j} - 1)| \to \infty$$

as $j \to \infty$. Hence $\lim_{j \to \infty} d_j = \infty$.

Problem 40.   (i) Consider the polynomial

$$p(x) = x^2 - sx + d, \quad s, d \in \mathbf{C}.$$

Find a 2×2 matrix A such that its characteristic polynomial is p.
(ii) Consider the polynomial

$$q(x) = -x^3 + sx^2 - qx + d, \quad s, q, d \in \mathbf{C}.$$

Find a 3×3 matrix B such that its characteristic polynomial is q.

Solution 40.   (i) We obtain

$$A = \begin{pmatrix} s & d \\ -1 & 0 \end{pmatrix}.$$

(ii) We obtain

$$B = \begin{pmatrix} s & q & d \\ -1 & 0 & 0 \\ 0 & -1 & 0 \end{pmatrix}.$$

Problem 41.   Calculate the eigenvalues of the 4×4 matrix

$$A = \begin{pmatrix} 1 & 0 & 0 & 1 \\ 0 & 1 & 1 & 0 \\ 0 & 1 & -1 & 0 \\ 1 & 0 & 0 & -1 \end{pmatrix}$$

by calculating the eigenvalues of A^2.

Solution 41.   The matrix A is symmetric over $\mathbf{R}$. Thus the eigenvalues of A are real. Now A^2 is the diagonal matrix

$$A^2 = \text{diag}(2\ 2\ 2\ 2)$$

with eigenvalue 2 (four times). Since $\text{tr}A = 0$ we obtain $\sqrt{2}, \sqrt{2}, -\sqrt{2}, -\sqrt{2}$ as the eigenvalues of A.

Chapter 5

Commutators and Anticommutators

Let A and B be $n \times n$ matrices. Then we define the *commutator* of A and B as

$$[A, B] := AB - BA.$$

For all $n \times n$ matrices A, B, C we have the *Jacobi identity*

$$[A, [B, C]] + [C, [A, B]] + [B, [C, A]] = 0_n$$

where 0_n is the $n \times n$ zero matrix. Since $\text{tr}(AB) = \text{tr}(BA)$ we have

$$\text{tr}([A, B]) = 0.$$

If

$$[A, B] = 0_n$$

we say that the matrices A and B *commute*. For example, if A and B are diagonal matrices then the commutator is the zero matrix 0_n.

Let A and B be $n \times n$ matrices. Then we define the *anticommutator* of A and B as

$$[A, B]_+ := AB + BA.$$

We have

$$\text{tr}([A, B]_+) = 2\text{tr}(AB).$$

The anticommutator plays a role for Fermi operators.

Problem 1.   Let A, B be $n \times n$ matrices. Assume that $[A, B] = 0_n$ and $[A, B]_+ = 0_n$. What can be said about AB and BA?

Solution 1.   Since $AB = BA$ and $AB = -BA$ we find

$$AB = 0_n, \qquad BA = 0_n.$$

Problem 2.   Let A and B be symmetric $n \times n$ matrices over **R**. Show that AB is symmetric if and only if A and B commute.

Solution 2.   Suppose that A and B commute, i.e. $AB = BA$. Then

$$(AB)^T = B^T A^T = BA = AB$$

and thus AB is symmetric. Suppose that AB is symmetric, i.e. $(AB)^T = AB$. Then $(AB)^T = B^T A^T = BA$. Hence $AB = BA$ and the matrices A and B commute.

Problem 3.   Let A and B be $n \times n$ matrices over **C**. Show that A and B commute if and only if $A - cI_n$ and $B - cI_n$ commute over every $c \in$ **C**.

Solution 3.   Suppose that A and B commute, i.e. $AB = BA$. Then

$$\begin{aligned}
(A - cI_n)(B - cI_n) &= AB - c(A + B) + c^2 I_n \\
&= BA - c(A + B) + c^2 I_n \\
&= (B - cI_n)(A - cI_n).
\end{aligned}$$

Thus $A - cI_n$ and $B - cI_n$ commute. Now suppose that $A - cI_n$ and $B - cI_n$ commute, i.e.

$$(A - cI_n)(B - cI_n) = (B - cI_n)(A - cI_n).$$

Thus

$$AB - c(A + B) + c^2 I_n = BA - c(A + B) + c^2 I_n.$$

Thus $AB = BA$.

Problem 4.   Consider the matrices

$$h = \begin{pmatrix} 1 & 0 \\ 0 & -1 \end{pmatrix}, \quad e = \begin{pmatrix} 0 & 1 \\ 0 & 0 \end{pmatrix}, \quad f = \begin{pmatrix} 0 & 0 \\ 1 & 0 \end{pmatrix}.$$

Find a nonzero 2×2 matrices A such that

$$[A, e] = 0_n, \quad [A, f] = 0_n, \quad [A, h] = 0_n.$$

Solution 4. From $[A, e] = 0_n$ we obtain $a_{21} = 0_n$ and $a_{11} = a_{22}$. From $[A, f] = 0_n$ we obtain $a_{12} = 0$ and $a_{11} = a_{22}$. Then it follows that $[A, h] = 0_n$. Thus the matrix is

$$A = \begin{pmatrix} a_{11} & 0 \\ 0 & a_{11} \end{pmatrix} = a_{11} \begin{pmatrix} 1 & 0 \\ 0 & 1 \end{pmatrix}$$

where $a_{11} \neq 0$. The matrix A plays the role as the *Casimir element* in a Lie algebra.

Problem 5. Can one find 2×2 matrices A and B such that

$$[A^2, B^2] = 0_n$$

while

$$[A, B] \neq 0_n ?$$

Solution 5. An example is the matrices

$$A = \begin{pmatrix} 0 & 1 \\ 0 & 0 \end{pmatrix}, \qquad B = \begin{pmatrix} 0 & 0 \\ 1 & 0 \end{pmatrix}.$$

Then A^2 and B^2 are the 2×2 zero matrix and

$$[A, B] = \begin{pmatrix} 1 & 0 \\ 0 & -1 \end{pmatrix}.$$

Problem 6. Let A, B, C, D be $n \times n$ matrices over $\mathbf{R}$. Assume that AB^T and CD^T are symmetric and $AD^T - BC^T = I_n$, where T denotes transpose. Show that

$$A^T D - C^T B = I_n.$$

Solution 6. From the assumptions we have

$$AB^T = (AB^T)^T = BA^T$$
$$CD^T = (CD^T)^T = DC^T$$
$$AD^T - BC^T = I_n.$$

Taking the transpose of the third equation we have

$$DA^T - CB^T = I_n.$$

These four equations can be written in the form of block matrices in the identity

$$\begin{pmatrix} A & B \\ C & D \end{pmatrix} \begin{pmatrix} D^T & -B^T \\ -C^T & A^T \end{pmatrix} = \begin{pmatrix} I_n & 0 \\ 0 & I_n \end{pmatrix}.$$

Thus the matrices are $(2n) \times (2n)$ marices. If X, Y are $m \times m$ matrices with $XY = I_m$, the identity matrix, then $Y = X^{-1}$ and $YX = I_m$ too. Applying this to our matrix equation with $m = 2n$ we obtain

$$\begin{pmatrix} D^T & -B^T \\ -C^T & A^T \end{pmatrix} \begin{pmatrix} A & B \\ C & D \end{pmatrix} = \begin{pmatrix} I_n & 0 \\ 0 & I_n \end{pmatrix}.$$

Equating the lower right blocks shows that $-C^T B + A^T D = I_n$.

Problem 7.   Let A, B, H be $n \times n$ matrices over $\mathbf{C}$ such that

$$[A, H] = 0_n, \qquad [B, H] = 0_n.$$

Find $[[A, B], H]$.

Solution 7.   Using the *Jacobi identity* for arbitrary $n \times n$ matrices X, Y, Z

$$[[X, Y], Z] + [[Z, X], Y] + [[Y, Z], X] = 0_n$$

we obtain

$$[[A, B], H] = 0_n.$$

Problem 8.   Let A, B be $n \times n$ matrices. Assume that A is invertible. Assume that $[A, B] = 0_n$. Can we conclude that $[A^{-1}, B] = 0_n$?

Solution 8.   From $[A, B] = 0_n$ we have

$$AB = BA.$$

Thus $B = A^{-1}BA$ and therefore $BA^{-1} = A^{-1}B$. It follows that

$$A^{-1}B - BA^{-1} = [A^{-1}, B] = 0_n.$$

Problem 9.   Let A and B be $n \times n$ hermitian matrices. Suppose that

$$A^2 = I_n, \qquad B^2 = I_n \tag{1}$$

and

$$[A, B]_+ \equiv AB + BA = 0_n \tag{2}$$

where 0_n is the $n \times n$ zero matrix. Let $\mathbf{x} \in \mathbf{C}^n$ be normalized, i.e., $\|\mathbf{x}\| = 1$. Here $\mathbf{x}$ is considered as a column vector.
(i) Show that

$$(\mathbf{x}^* A \mathbf{x})^2 + (\mathbf{x}^* B \mathbf{x})^2 \leq 1. \tag{3}$$

(ii) Give an example for the matrices A and B.

Solution 9. (i) Let $a, b \in \mathbf{R}$ and let $r^2 := a^2 + b^2$. The matrix

$$C = aA + bB$$

is again hermitian. Then

$$C^2 = a^2 A^2 + abAB + baBA + b^2 B^2 .$$

Using the properties (1) and (2) we find

$$C^2 = a^2 I_n + b^2 I_n = r^2 I_n .$$

Therefore

$$(\mathbf{x}^* C^2 \mathbf{x}) = r^2$$

and

$$-r \le a(\mathbf{x}^* A \mathbf{x}) + b(\mathbf{x}^* B \mathbf{x}) \le r .$$

Let

$$a = \mathbf{x}^* A \mathbf{x}, \qquad b = \mathbf{x}^* B \mathbf{x}$$

then

$$a^2 + b^2 \le r$$

or $r^2 \le r$. This implies $r \le 1$ and $r^2 \le 1$ from which (3) follows.

(ii) An example is $A = \sigma_x$ and $B = \sigma_y$ since $\sigma_x^2 = I_2$, $\sigma_y^2 = I_2$ and $\sigma_x \sigma_y + \sigma_y \sigma_x = 0_2$, where σ_x, σ_y, σ_z are the Pauli spin matrices.

Problem 10. Let A and B be $n \times n$ hermitian matrices. Suppose that

$$A^2 = A, \qquad B^2 = B \tag{1}$$

and

$$[A, B]_+ \equiv AB + BA = 0_n \tag{2}$$

where 0_n is the $n \times n$ zero matrix. Let $\mathbf{x} \in \mathbf{C}^n$ be normalized, i.e., $\|\mathbf{x}\| = 1$. Here $\mathbf{x}$ is considered as a column vector. Show that

$$(\mathbf{x}^* A \mathbf{x})^2 + (\mathbf{x}^* B \mathbf{x})^2 \le 1 . \tag{3}$$

Solution 10. For an arbitrary $n \times n$ hermitian matrix M we have

$$0 \le (\mathbf{x}^* (M - (\mathbf{x}^* M \mathbf{x}) I_n)^2 \mathbf{x}) = (\mathbf{x}^* (M^2 - 2(\mathbf{x}^* M \mathbf{x}) M + (\mathbf{x}^* M \mathbf{x})^2 I_n) \mathbf{x})$$
$$= (\mathbf{x}^* M^2 \mathbf{x}) - 2(\mathbf{x}^* M \mathbf{x})^2 + (\mathbf{x}^* M \mathbf{x})^2 = (\mathbf{x}^* M^2 \mathbf{x}) - (\mathbf{x}^* M \mathbf{x})^2 .$$

Thus
$$0 \leq (\mathbf{x}^* M^2 \mathbf{x}) - (\mathbf{x}^* M \mathbf{x})^2$$

or
$$(\mathbf{x}^* M \mathbf{x})^2 \leq (\mathbf{x}^* M^2 \mathbf{x}).$$

Thus for $A = M$ we have using (1)
$$(\mathbf{x}^* A \mathbf{x})^2 \leq \mathbf{x}^* A \mathbf{x}$$

and therefore
$$0 \leq (\mathbf{x}^* A \mathbf{x}) \leq 1.$$

Similarly
$$0 \leq (\mathbf{x}^* B \mathbf{x}) \leq 1.$$

Let $a, b \in \mathbf{R}$, $r^2 := a^2 + b^2$ and
$$C := aA + bB.$$

Then
$$C^2 = a^2 A^2 + b^2 B^2 + abAB + baBA.$$

Using (1) and (2) we arrive at
$$C^2 = a^2 A + b^2 B.$$

Thus
$$(\mathbf{x}^* C \mathbf{x})^2 \leq (\mathbf{x}^* C^2 \mathbf{x}) \leq a^2 + b^2.$$

Let
$$a := \mathbf{x}^* A \mathbf{x}, \qquad b := \mathbf{x}^* B \mathbf{x}$$

then
$$\mathbf{x}^* C \mathbf{x} = a^2 + b^2 = r^2$$

and therefore $(r^2)^2 \leq r^2$ which implies that $r^2 \leq 1$ and thus (3) follows.

Problem 11. Let A, B be skew-hermitian matrices over $\mathbf{C}$, i.e. $A^* = -A$, $B^* = -B$. Is the commutator of A and B again skew-hermitian?

Solution 11. The answer is yes. We have
$$\begin{aligned}
([A, B])^* &= (AB - BA)^* \\
&= B^* A^* - A^* B^* \\
&= BA - AB \\
&= -([A, B]).
\end{aligned}$$

Problem 12. Let A, B be $n \times n$ matrices over $\mathbf{C}$. Let S be an invertible $n \times n$ matrix over $\mathbf{C}$ with

$$\tilde{A} = S^{-1}AS, \qquad \tilde{B} = S^{-1}BS.$$

Show that

$$[\tilde{A}, \tilde{B}] = S^{-1}[A, B]S.$$

Solution 12. Since $SS^{-1} = I_n$ we have

$$
\begin{aligned}
[\tilde{A}, \tilde{B}] &= [S^{-1}AS, S^{-1}BS] \\
&= S^{-1}ASS^{-1}BS - S^{-1}BSS^{-1}AS \\
&= S^{-1}ABS - S^{-1}BAS \\
&= S^{-1}(AB - BA)S \\
&= S^{-1}[A, B]S.
\end{aligned}
$$

Problem 13. Can we find $n \times n$ matrices A, B over $\mathbf{C}$ such that

$$[A, B] = I_n \tag{1}$$

where I_n denotes the identity matrix?

Solution 13. Since $\mathrm{tr}(XY) = \mathrm{tr}(YX)$ for any $n \times n$ matrices X, Y we obtain $\mathrm{tr}([A, B]) = 0$. However for the right-hand side of (1) we find $\mathrm{tr}I_n = n$. Thus we have a contradiction and no such matrices exist.

We can find unbounded infinite-dimensional matrices X, Y such that $[X, Y] = I$, where I is the infinite-dimensional unit matrix.

Problem 14. Can we find 2×2 matrices A and B of the form

$$A = \begin{pmatrix} 0 & a_{12} \\ a_{21} & 0 \end{pmatrix}, \qquad B = \begin{pmatrix} 0 & b_{12} \\ b_{21} & 0 \end{pmatrix}$$

and singular (i.e. $\det A = 0$ and $\det B = 0$) such that

$$[A, B]_+ = I_2.$$

Solution 14. We have

$$AB + BA = \begin{pmatrix} a_{12}b_{21} + a_{21}b_{12} & 0 \\ 0 & a_{12}b_{21} + a_{21}b_{12} \end{pmatrix} = \begin{pmatrix} 1 & 0 \\ 0 & 1 \end{pmatrix}.$$

Thus we obtain the equation

$$a_{12}b_{21} + a_{21}b_{12} = 1.$$

Since A and B are singular we obtain the two solutions

$$a_{12} = b_{21} = 1, \qquad a_{21} = b_{12} = 0$$

and

$$a_{12} = b_{21} = 0, \qquad a_{21} = b_{12} = 1.$$

Problem 15.   Let A be an $n \times n$ hermitian matrix over **C**. Assume that the eigenvalues of A, $\lambda_1, \lambda_2, \ldots, \lambda_n$ are nondegenerate and that the normalized eigenvectors $\mathbf{v}_j$ ($j = 1, 2, \ldots, n$) of A form an orthonormal basis in $\mathbf{C}^n$. Let B be an $n \times n$ matrix over **C**. Assume that $[A, B] = 0_n$, i.e., A and B commute. Show that

$$\mathbf{v}_k^* B \mathbf{v}_j = 0 \quad \text{for} \quad k \neq j. \tag{1}$$

Solution 15.   From $AB = BA$ it follows that

$$\mathbf{v}_k^*(AB\mathbf{v}_j) = \mathbf{v}_k^*(BA\mathbf{v}_j).$$

Note that the eigenvalues of a hermitian matrix are real. Since $A\mathbf{v}_j = \lambda_j \mathbf{v}_j$ and $\mathbf{v}_k^* A = \lambda_k \mathbf{v}_k^*$ we obtain

$$\lambda_k(\mathbf{v}_k^* B \mathbf{v}_j) = \lambda_j(\mathbf{v}_k^* B \mathbf{v}_j).$$

Consequently

$$(\lambda_k - \lambda_j)(\mathbf{v}_k^* B \mathbf{v}_j) = 0.$$

Since $\lambda_k \neq \lambda_j$ equation (1) follows.

Problem 16.   Let A, B be hermitian $n \times n$ matrices. Assume they have the same set of eigenvectors

$$A\mathbf{v}_j = \lambda_j \mathbf{v}_j, \quad B\mathbf{v}_j = \mu_j \mathbf{v}_j, \quad j = 1, 2, \ldots, n$$

and that the normalized eigenvectors form an orthonormal basis in $\mathbf{C}^n$. Show that

$$[A, B] = 0_n. \tag{1}$$

Solution 16.   Any vector $\mathbf{v}$ in $\mathbf{C}^n$ can be written as

$$\mathbf{v} = \sum_{j=1}^{n} (\mathbf{v}^* \mathbf{v}_j)\mathbf{v}_j.$$

It follows that

$$[A, B]\mathbf{v} = (AB - BA) \sum_{j=1}^{n} (\mathbf{v}^* \mathbf{v}_j) \mathbf{v}_j$$

$$= \sum_{j=1}^{n} (\mathbf{v}^* \mathbf{v}_j) AB\mathbf{v}_j - \sum_{j=1}^{n} (\mathbf{v}^* \mathbf{v}_j) BA\mathbf{v}_j$$

$$= \sum_{j=1}^{n} (\mathbf{v}^* \mathbf{v}_j) A\mu_j \mathbf{v}_j - \sum_{j=1}^{n} (\mathbf{v}^* \mathbf{v}_j) B\lambda_j \mathbf{v}_j$$

$$= \sum_{j=1}^{n} (\mathbf{v}^* \mathbf{v}_j) \lambda_j \mu_j \mathbf{v}_j - \sum_{j=1}^{n} (\mathbf{v}^* \mathbf{v}_j) \mu_j \lambda_j \mathbf{v}_j$$

$$= 0.$$

Since this is true for an arbitrary vector $\mathbf{v}$ in $\mathbf{C}^n$ equation (1) follows.

Problem 17. Let A, B be $n \times n$ matrices. Then we have the expansion

$$e^A B e^{-A} = B + [A, B] + \frac{1}{2!}[A, [A, B]] + \frac{1}{3!}[A, [A, [A, B]]] + \cdots$$

(i) Assume that $[A, B] = A$. Calculate $e^A B e^{-A}$.
(ii) Assume that $[A, B] = B$. Calculate $e^A B e^{-A}$.

Solution 17. (i) From $[A, B] = A$ we obtain

$$[A, [A, B]] = [A, A] = 0_n.$$

Thus

$$e^A B e^{-A} = B + A.$$

(ii) From $[A, B] = B$ we obtain

$$[A, [A, B]] = B, \qquad [A, [A, [A, B]]] = B$$

etc. Thus

$$e^A B e^{-A} = B + B + \frac{1}{2!}B + \frac{1}{3!}B + \cdots = B\left(1 + 1 + \frac{1}{2!} + \frac{1}{3!} + \cdots\right) = eB.$$

Problem 18. Let A be an arbitrary $n \times n$ matrix over $\mathbf{C}$ with $\text{tr}A = 0$. Show that A can be written as commutator, i.e., there are $n \times n$ matrices X and Y such that $A = [X, Y]$.

Solution 18.   The statement is obviously valid if A is 1×1 or a larger $n \times n$ zero matrix. Therefore assume that A is a nonzero $n \times n$ matrix of dimension larger than 1. We use induction. We assume the desired inference valid for all matrices of dimensions smaller than A's with trace zero. Since $\text{tr}A = 0$ the matrix A cannot be a nonzero scalar multiple of I_n. Thus there is some invertible matrix R such that

$$R^{-1}AR = \begin{pmatrix} 0 & d^T \\ c & B \end{pmatrix}$$

with $\text{tr}B = 0$. The induction hypothesis implies that B is a commutator. Thus $R^{-1}AR = XY - YX$ is also a commutator for some $n \times n$ matrices X and Y. It follows that

$$A = (RXR^{-1})(RYR^{-1}) - (RYR^{-1})(RXR^{-1})$$

must also be a commutator.

Problem 19.   (i) Let A, B be $n \times n$ matrices over $\mathbf{C}$ with $[A, B] = 0_n$. Calculate

$$[A + cI_n, B + cI_n]$$

where $c \in \mathbf{C}$ and I_n is the $n \times n$ identity matrix.
(ii) Let $\mathbf{x}$ be an eigenvector of the $n \times n$ matrix A with eigenvalue λ. Show that $\mathbf{x}$ is also an eigenvector of $A + cI_n$, where $c \in \mathbf{C}$.

Solution 19.   (i) We obtain

$$[A + cI_n, B + cI_n] = [A, B] + c[A, I_n] + c[I_n, B] + c^2[I_n, I_n] = 0.$$

(ii) We have

$$(A + cI_n)\mathbf{x} = A\mathbf{x} + cI_n\mathbf{x}$$
$$= \lambda\mathbf{x} + c\mathbf{x}$$
$$= (\lambda + c)\mathbf{x}.$$

Thus $\mathbf{x}$ is an eigenvector of $A + cI_n$ with eigenvalue $\lambda + c$.

Chapter 6

Decomposition of Matrices

A matrix decomposition (or matrix factorization) is the right-hand side matrix product $A = F_1 F_2 \cdots F_n$ for an input matrix A. The number of factor matrices $F_1, F_2, \ldots, F_n$ depends on the chosen decomposition. In most cases $n = 2$ or $n = 3$.

The most common decompositions are:

1) *LU-decomposition.* A square matrix A is factorized into a product of a lower triangular matrix, L, and an upper triangular matrix U, i.e. $A = LU$.

2) *QR-decomposition.* An $n \times m$ matrix with linearly independent columns is factorized as $A = QR$, where Q is an $n \times m$ matrix with orthonormal columns and R is an invertible $m \times m$ upper triangular matrix.

3) The *polar decomposition* of $A \in \mathbf{C}^{n \times n}$ factors A as the product $A = UH$, where U is unitary and H is hermitian positive semi-definite. The hermitian factor is always unique and can be expressed as $(A^*A)^{1/2}$, and the unitary factor is unique if A is nonsingular.

4) In the *singular value decomposition* an $m \times n$ matrix can be written as $A = U\Sigma V^T$, where U is an $m \times m$ orthogonal matrix, V is an $n \times n$ orthogonal matrix, Σ is an $m \times n$ diagonal matrix with nonnegative entries and the superscript ^{T} denotes the transpose.

Other important decompositions are the cosine-sine decomposition, the Cholesky decomposition, the Jordan decomposition, and the Iwasawa decomposition.

Problem 1. Find the LU-decomposition of the 3×3 matrix

$$A = \begin{pmatrix} 3 & 6 & -9 \\ 2 & 5 & -3 \\ -4 & 1 & 10 \end{pmatrix}.$$

The triangular matrices L and U are not uniquely determined by the matrix equation $A = LU$. These two matrices together contain $n^2 + n$ unknown elements. Thus when comparing elements on the left- and right-hand side of $A = LU$ we have n^2 equations and $n^2 + n$ unknowns. We require a further n conditions to uniquely determine the matrices. There are three additional sets of n conditions that are commonly used. These are *Doolittle's method* with $\ell_{jj} = 1$, $j = 1, 2, \ldots, n$; *Choleski's method* with $\ell_{jj} = u_{jj}$, $j = 1, 2, \ldots, n$; *Crout's method* with $u_{jj} = 1$, $j = 1, 2, \ldots, n$. Apply Crout's method.

Solution 1. From

$$L = \begin{pmatrix} \ell_{11} & 0 & 0 \\ \ell_{21} & \ell_{22} & 0 \\ \ell_{31} & \ell_{32} & \ell_{33} \end{pmatrix}, \qquad U = \begin{pmatrix} 1 & u_{12} & u_{13} \\ 0 & 1 & u_{23} \\ 0 & 0 & 1 \end{pmatrix}$$

and $A = LU$ we obtain the 9 equations

$$\ell_{11} = 3$$
$$\ell_{11} u_{12} = 6$$
$$\ell_{11} u_{13} = -9$$
$$\ell_{21} = 2$$
$$\ell_{21} u_{12} + \ell_{22} = 5$$
$$\ell_{21} u_{13} + \ell_{22} u_{23} = -3$$
$$\ell_{31} = -4$$
$$\ell_{31} u_{12} + \ell_{32} = 1$$
$$\ell_{31} u_{13} + \ell_{32} u_{23} + \ell_{33} = 20$$

with the solution

$$\ell_{11} = 3, \quad \ell_{21} = 2, \quad \ell_{22} = 1, \quad \ell_{31} = -4, \quad \ell_{32} = 9, \quad \ell_{33} = -29$$

$$u_{12} = 2, \quad u_{13} = -3, \quad u_{23} = 3.$$

Thus we obtain

$$L = \begin{pmatrix} 3 & 0 & 0 \\ 2 & 1 & 0 \\ -4 & 9 & -29 \end{pmatrix}, \qquad U = \begin{pmatrix} 1 & 2 & -3 \\ 0 & 1 & 3 \\ 0 & 0 & 1 \end{pmatrix}.$$

Problem 2. Find the QR-decomposition of the 3×3 matrix

$$A = \begin{pmatrix} 2 & 1 & 3 \\ -1 & 0 & 7 \\ 0 & -1 & -1 \end{pmatrix}.$$

Solution 2. The columns of A are

$$\mathbf{c}_1 = \begin{pmatrix} 2 \\ -1 \\ 0 \end{pmatrix}, \quad \mathbf{c}_2 = \begin{pmatrix} -1 \\ 0 \\ -1 \end{pmatrix}, \quad \mathbf{c}_3 = \begin{pmatrix} 3 \\ 7 \\ -1 \end{pmatrix}.$$

They are linearly independent. Applying the *Gram-Schmidt orthonormalization process* we obtain

$$\mathbf{u}_1 = \begin{pmatrix} 2/\sqrt{5} \\ -1/\sqrt{5} \\ 0 \end{pmatrix}, \quad \mathbf{u}_2 = \begin{pmatrix} 1/\sqrt{30} \\ 2/\sqrt{30} \\ -5/\sqrt{30} \end{pmatrix}, \quad \mathbf{u}_3 = \begin{pmatrix} 1/\sqrt{6} \\ 2/\sqrt{6} \\ 1/\sqrt{6} \end{pmatrix}.$$

Thus we obtain

$$Q = \begin{pmatrix} 2/\sqrt{5} & 1/\sqrt{30} & 1/\sqrt{6} \\ -1/\sqrt{5} & 2/\sqrt{30} & 2/\sqrt{6} \\ 0 & -5/\sqrt{30} & 1/\sqrt{6} \end{pmatrix}$$

and

$$R = \begin{pmatrix} \sqrt{5} & 2/\sqrt{5} & -1/\sqrt{5} \\ 0 & 6/\sqrt{30} & 22/\sqrt{30} \\ 0 & 0 & 16/\sqrt{6} \end{pmatrix}.$$

Problem 3. Consider a square non-singular square matrix A over $\mathbf{C}$, i.e. A^{-1} exists. The *polar decomposition theorem* states that A can be written as

$$A = UP$$

where U is a unitary matrix and P is a hermitian positive definite matrix. Show that A has a unique polar decomposition.

Solution 3. Since A is invertible, so are A^* and A^*A. The positive square root P of A^*A is also invertible. Set $U := AP^{-1}$. Then U is invertible and

$$U^*U = P^{-1}A^*AP^{-1} = P^{-1}P^2P^{-1} = I$$

so that U is unitary. Since P is invertible, it is obvious that AP^{-1} is the only possible choice for U.

Problem 4. Let A be an arbitrary $m \times n$ matrix over $\mathbf{R}$, i.e., $A \in \mathbf{R}^{m \times n}$. Then A can be written as

$$A = U \Sigma V^T$$

where U is an $m \times m$ orthogonal matrix, V is an $n \times n$ orthogonal matrix, Σ is an $m \times n$ diagonal matrix with nonnegative entries and the superscript T denotes the transpose. This is called the *singular value decomposition*. An algorithm to find the singular value decomposition is as follows.

1) Find the eigenvalues λ_j $(j = 1, 2, \dots, n)$ of the $n \times n$ matrix $A^T A$. Arrange the eigenvalues $\lambda_1, \lambda_2, \dots, \lambda_n$ in descending order.

2) Find the number of nonzero eigenvalues of the matrix $A^T A$. We call this number r.

3) Find the orthogonal eigenvectors $\mathbf{v}_j$ of the matrix $A^T A$ corresponding to the obtained eigenvalues, and arrange them in the same order to form the column-vectors of the $n \times n$ matrix V.

4) Form an $m \times n$ diagonal matrix Σ placing on the leading diagonal of it the square root $\sigma_j := \sqrt{\lambda_j}$ of $p = \min(m, n)$ first eigenvalues of the matrix $A^T A$ found in 1) in descending order.

5) Find the first r column vectors of the $m \times m$ matrix U

$$\mathbf{u}_j = \frac{1}{\sigma_j} A \mathbf{v}_j, \quad j = 1, 2, \dots, r.$$

6) Add to the matrix U the rest of the $m - r$ vectors using the Gram-Schmidt orthogonalization process.

We have

$$A \mathbf{v}_j = \sigma_j \mathbf{u}_j, \qquad A^T \mathbf{u}_j = \sigma_j \mathbf{v}_j$$

and therefore

$$A^T A \mathbf{v}_j = \sigma_j^2 \mathbf{v}_j, \qquad A A^T \mathbf{u}_j = \sigma_j^2 \mathbf{u}_j.$$

Apply the algorithm to the matrix

$$A = \begin{pmatrix} 0.96 & 1.72 \\ 2.28 & 0.96 \end{pmatrix}.$$

Solution 4. 1) We find

$$A^T A = \begin{pmatrix} 6.12 & 3.84 \\ 3.84 & 3.88 \end{pmatrix}.$$

The eigenvalues are (arranged in descending order) $\lambda_1 = 9$ and $\lambda_2 = 1$.

2) The number of nonzero eigenvalues is $r = 2$.

3) The orthonormal normalized eigenvectors of the matrix $A^T A$, corresponding to the eigenvalues λ_1 and λ_2 are given by

$$\mathbf{v}_1 = \begin{pmatrix} 0.8 \\ 0.6 \end{pmatrix}, \qquad \mathbf{v}_2 = \begin{pmatrix} 0.6 \\ -0.8 \end{pmatrix}.$$

We obtain the 2×2 matrix V (V^T follows by taking the transpose)

$$V = (\mathbf{v}_1 \ \mathbf{v}_2) = \begin{pmatrix} 0.8 & 0.6 \\ 0.6 & -0.8 \end{pmatrix}.$$

4) From the eigenvalues we find the singular matrix taking the square roots of the eigenvalues

$$\Sigma = \begin{pmatrix} 3 & 0 \\ 0 & 1 \end{pmatrix}.$$

5) Next we find two column vectors of the 2×2 matrix U. Using the equation given above we find ($\sigma_1 = 3, \sigma_2 = 1$)

$$\mathbf{u}_1 = \frac{1}{\sigma_1} A \mathbf{v}_1 = \begin{pmatrix} 0.6 \\ 0.8 \end{pmatrix}, \qquad \mathbf{u}_2 = \frac{1}{\sigma_2} A \mathbf{v}_2 = \begin{pmatrix} -0.8 \\ 0.6 \end{pmatrix}.$$

It follows that

$$U = (\mathbf{u}_1 \ \mathbf{u}_2) = \begin{pmatrix} 0.6 & -0.8 \\ 0.8 & 0.6 \end{pmatrix}.$$

Thus we have found the singular value decomposition of the matrix A

$$A = U \Sigma V^T = \begin{pmatrix} 0.6 & -0.8 \\ 0.8 & 0.6 \end{pmatrix} \begin{pmatrix} 3 & 0 \\ 0 & 1 \end{pmatrix} \begin{pmatrix} 0.8 & 0.6 \\ 0.6 & -0.8 \end{pmatrix}^T.$$

Problem 5. Find the singular value decomposition $A = U \Sigma V^T$ of the matrix (row vector) $A = (2 \ 1 \ -2)$.

Solution 5. First we find the eigenvalues of the 3×3 matrix $A^T A$, where ^{T} denotes transpose. We have

$$A^T A = \begin{pmatrix} 4 & 2 & -4 \\ 2 & 1 & -2 \\ -4 & -2 & 4 \end{pmatrix}.$$

The eigenvalues are given by $\lambda_1 = 9$, $\lambda_2 = 0$, $\lambda_3 = 0$. Thus the eigenvalue 0 is degenerate. Next we find the number r of nonzero eigenvalues which is obviously $r = 1$. Now we calculate the normalized eigenvectors. For the eigenvalue $\lambda_1 = 9$ we find

$$\mathbf{v}_1 = (-2/3 \; -1/3 \; 2/3)^T .$$

To find the two normalized eigenvectors for the eigenvalue 0 we have to apply the Gram-Schmidt orthogonalization process. Then we find

$$\mathbf{v}_2 = \begin{pmatrix} -\sqrt{5}/5 \\ 2\sqrt{5}/5 \\ 0 \end{pmatrix}, \qquad \mathbf{v}_3 = \begin{pmatrix} 4\sqrt{5}/15 \\ 2\sqrt{5}/15 \\ 5\sqrt{5}/15 \end{pmatrix} .$$

This provides the orthonormal matrix

$$V = \begin{pmatrix} -2/3 & -\sqrt{5}/5 & 4\sqrt{5}/15 \\ -1/3 & 2\sqrt{5}/5 & 2\sqrt{5}/15 \\ 2/3 & 0 & \sqrt{5}/3 \end{pmatrix} .$$

Next we form the singular value matrix

$$\Sigma = (3 \; 0 \; 0) .$$

Finally we calculate the unique column-vector of the matrix U

$$\mathbf{u}_1 = \frac{1}{3} A \mathbf{v}_1 = \frac{1}{3}(2 \; 1 \; -2) \begin{pmatrix} -2/3 \\ -1/3 \\ 2/3 \end{pmatrix} = (-1) .$$

Hence the singular value decomposition of the matrix A is

$$A = U\Sigma V^T = (-1)(3 \; 0 \; 0) \begin{pmatrix} -2/3 & -1/3 & 2/3 \\ \sqrt{5}/5 & 2\sqrt{5}/5 & 0 \\ 4\sqrt{5}/15 & 2\sqrt{5}/15 & \sqrt{5}/3 \end{pmatrix} .$$

Problem 6.   Any unitary $2^n \times 2^n$ matrix U can be decomposed as

$$U = \begin{pmatrix} U_1 & 0 \\ 0 & U_2 \end{pmatrix} \begin{pmatrix} C & S \\ -S & C \end{pmatrix} \begin{pmatrix} U_3 & 0 \\ 0 & U_4 \end{pmatrix}$$

where U_1, U_2, U_3, U_4 are $2^{n-1} \times 2^{n-1}$ unitary matrices and C and S are the $2^{n-1} \times 2^{n-1}$ diagonal matrices

$$C = \operatorname{diag}(\cos \alpha_1, \cos \alpha_2, \ldots, \cos \alpha_{2^n/2})$$
$$S = \operatorname{diag}(\sin \alpha_1, \sin \alpha_2, \ldots, \sin \alpha_{2^n/2})$$

where $\alpha_j \in \mathbf{R}$. This decomposition is called *cosine-sine decomposition*.

Consider the unitary 2×2 matrix

$$U = \begin{pmatrix} 0 & i \\ -i & 0 \end{pmatrix}.$$

Show that U can be written as

$$U = \begin{pmatrix} u_1 & 0 \\ 0 & u_2 \end{pmatrix} \begin{pmatrix} \cos\alpha & \sin\alpha \\ -\sin\alpha & \cos\alpha \end{pmatrix} \begin{pmatrix} u_3 & 0 \\ 0 & u_4 \end{pmatrix}$$

where $\alpha \in \mathbf{R}$ and $u_1, u_2, u_3, u_4 \in U(1)$ (i.e., u_1, u_2, u_3, u_4 are complex numbers with length 1). Find α, u_1, u_2, u_3, u_4.

Solution 6. Matrix multiplication yields

$$\begin{pmatrix} 0 & i \\ -i & 0 \end{pmatrix} = \begin{pmatrix} u_1 u_3 \cos\alpha & u_1 u_4 \sin\alpha \\ -u_2 u_3 \sin\alpha & u_2 u_4 \cos\alpha \end{pmatrix}.$$

Since $u_1, u_2, u_3, u_4 \neq 0$ we obtain $\cos\alpha = 0$. We select the solution $\alpha = \pi/2$. Since $\sin(\pi/2) = 1$ it follows that

$$u_1 u_4 = i, \qquad u_2 u_3 = i.$$

Thus we can select the solution $u_1 = u_2 = 1$, $u_3 = u_4 = i$.

Problem 7. (i) Find the *cosine-sine decomposition* of the unitary matrix

$$U = \begin{pmatrix} 0 & 1 \\ 1 & 0 \end{pmatrix}.$$

(ii) Use the result from (i) to find a 2×2 hermitian matrix K such that $U = \exp(iK)$.

Solution 7. (i) We have $(\alpha \in \mathbf{R})$

$$\begin{pmatrix} 0 & 1 \\ 1 & 0 \end{pmatrix} = \begin{pmatrix} u_1 & 0 \\ 0 & u_2 \end{pmatrix} \begin{pmatrix} \cos\alpha & \sin\alpha \\ -\sin\alpha & \cos\alpha \end{pmatrix} \begin{pmatrix} u_3 & 0 \\ 0 & u_4 \end{pmatrix}$$

where $u_1, u_2, u_3, u_4 \in \mathbf{C}$ with $|u_1| = |u_2| = |u_3| = |u_4| = 1$. Matrix multiplication yields

$$\begin{pmatrix} 0 & 1 \\ 1 & 0 \end{pmatrix} = \begin{pmatrix} u_1 u_3 \cos\alpha & u_1 u_4 \sin\alpha \\ -u_2 u_3 \sin\alpha & u_2 u_4 \cos\alpha \end{pmatrix}.$$

Since $u_1, u_2, u_3, u_4 \neq 0$ we obtain $\cos\alpha = 0$. We select the solution $\alpha = \pi/2$. Thus $\sin(\pi/2) = 1$ and

$$u_1 u_4 = 1, \qquad -u_2 u_3 = 1.$$

We select the solution $u_1 = u_4 = 1$, $u_2 = u_3 = i$. Thus we obtain the decomposition

$$\begin{pmatrix} 0 & 1 \\ 1 & 0 \end{pmatrix} = \begin{pmatrix} 1 & 0 \\ 0 & i \end{pmatrix} \begin{pmatrix} 0 & 1 \\ -1 & 0 \end{pmatrix} \begin{pmatrix} i & 0 \\ 0 & 1 \end{pmatrix}.$$

(ii) From (i) we obtain

$$\begin{pmatrix} 0 & 1 \\ 1 & 0 \end{pmatrix} = \begin{pmatrix} 1 & 0 \\ 0 & i \end{pmatrix} \begin{pmatrix} 0 & 1 \\ -1 & 0 \end{pmatrix} \begin{pmatrix} i & 0 \\ 0 & 1 \end{pmatrix} = e^{iK}.$$

The first matrix on the right-hand side is the unitary matrix

$$V = \begin{pmatrix} 1 & 0 \\ 0 & i \end{pmatrix}.$$

Therefore

$$V^* = \begin{pmatrix} 1 & 0 \\ 0 & -i \end{pmatrix}.$$

Thus

$$\begin{pmatrix} i & 0 \\ 0 & 1 \end{pmatrix} = iV^* = iV^{-1}$$

and

$$iV \begin{pmatrix} 0 & 1 \\ -1 & 0 \end{pmatrix} V^* = e^{i\pi I_2/2} V \begin{pmatrix} 0 & 1 \\ -1 & 0 \end{pmatrix} V^* = e^{iK}.$$

It follows that

$$V \begin{pmatrix} 0 & 1 \\ -1 & 0 \end{pmatrix} V^* = e^{i(K - \pi I_2/2)}$$

or

$$\begin{pmatrix} 0 & 1 \\ -1 & 0 \end{pmatrix} = V^* e^{i(K - \pi I_2/2)} V = e^{iV^*(K - \pi I_2/2)V}.$$

For $\alpha = \pi/2$ we have

$$\begin{pmatrix} 0 & 1 \\ 1 & 0 \end{pmatrix} = \begin{pmatrix} \cos(\pi/2) & \sin(\pi/2) \\ -\sin(\pi/2) & \cos(\pi/2) \end{pmatrix} = \exp\left(\alpha \begin{pmatrix} 0 & 1 \\ -1 & 0 \end{pmatrix} \right)\Big|_{\alpha=\pi/2}.$$

Comparing the exponents yields

$$\begin{pmatrix} 0 & \pi/2 \\ -\pi/2 & 0 \end{pmatrix} = iV^*(K - \pi I_2/2)V.$$

Since K is a hermitian matrix we can write

$$K = \begin{pmatrix} a & b \\ \bar{b} & d \end{pmatrix}, \qquad a, d \in \mathbf{R}.$$

Thus

$$
\begin{pmatrix} 0 & \pi/2 \\ -\pi/2 & 0 \end{pmatrix} = i \begin{pmatrix} 1 & 0 \\ 0 & -i \end{pmatrix} \begin{pmatrix} a & b \\ \bar{b} & d \end{pmatrix} \begin{pmatrix} 1 & 0 \\ 0 & i \end{pmatrix} - i\frac{\pi}{2} \begin{pmatrix} 1 & 0 \\ 0 & 1 \end{pmatrix}
$$
$$
= i \begin{pmatrix} a - \pi/2 & ib \\ -i\bar{b} & d - \pi/2 \end{pmatrix} .
$$

We obtain $a = \pi/2, b = -\pi/2$. Finally

$$
K = \begin{pmatrix} \pi/2 & -\pi/2 \\ -\pi/2 & \pi/2 \end{pmatrix} = \frac{\pi}{2} \begin{pmatrix} 1 & -1 \\ -1 & 1 \end{pmatrix} .
$$

Problem 8. (i) Find the *cosine-sine decomposition* of the unitary matrix (Hadamard matrix)

$$
U = \frac{1}{\sqrt{2}} \begin{pmatrix} 1 & 1 \\ 1 & -1 \end{pmatrix} .
$$

Solution 8. (i) We have ($\alpha \in \mathbf{R}$)

$$
\frac{1}{\sqrt{2}} \begin{pmatrix} 1 & 1 \\ 1 & -1 \end{pmatrix} = \begin{pmatrix} u_1 & 0 \\ 0 & u_2 \end{pmatrix} \begin{pmatrix} \cos\alpha & \sin\alpha \\ -\sin\alpha & \cos\alpha \end{pmatrix} \begin{pmatrix} u_3 & 0 \\ 0 & u_4 \end{pmatrix}
$$

where $u_1, u_2, u_3, u_4 \in \mathbf{C}$ with $|u_1| = |u_2| = |u_3| = |u_4| = 1$. Matrix multiplication yields

$$
\frac{1}{\sqrt{2}} \begin{pmatrix} 1 & 1 \\ 1 & -1 \end{pmatrix} = \begin{pmatrix} u_1 u_3 \cos\alpha & u_1 u_4 \sin\alpha \\ -u_2 u_3 \sin\alpha & u_2 u_4 \cos\alpha \end{pmatrix} .
$$

Thus we obtain the four equations

$$
\frac{1}{\sqrt{2}} = u_1 u_3 \cos\alpha, \qquad \frac{1}{\sqrt{2}} = -u_2 u_4 \cos\alpha
$$

$$
\frac{1}{\sqrt{2}} = u_1 u_4 \sin\alpha, \qquad \frac{1}{\sqrt{2}} = -u_2 u_3 \sin\alpha
$$

with a solution $\alpha = \pi/4$ and

$$
u_1 = u_3 = u_4 = 1, \qquad u_2 = -1 .
$$

Thus we obtain the decomposition

$$
\frac{1}{\sqrt{2}} \begin{pmatrix} 1 & 1 \\ 1 & -1 \end{pmatrix} = \begin{pmatrix} 1 & 0 \\ 0 & -1 \end{pmatrix} \begin{pmatrix} 1/\sqrt{2} & 1/\sqrt{2} \\ -1/\sqrt{2} & 1/\sqrt{2} \end{pmatrix} \begin{pmatrix} 1 & 0 \\ 0 & 1 \end{pmatrix} .
$$

Problem 9.   For any $n \times n$ matrix A there exists an $n \times n$ unitary matrix $(U^* = U^{-1})$ such that

$$U^*AU = T \tag{1}$$

where T is an $n \times n$ matrix in upper triangular form. Equation (1) is called a *Schur decomposition*. The diagonal elements of T are the eigenvalues of A. Note that such a decomposition is not unique. An iterative algorithm to find a Schur decomposition for an $n \times n$ matrix is as follows.

It generates at each step matrices U_k and T_k $(k = 1, 2, \ldots, n - 1)$ with the properties: each U_k is unitary, and each T_k has only zeros below its main diagonal in its first k columns. T_{n-1} is in upper triangular form, and $U = U_1 U_2 \cdots U_{n-1}$ is the unitary matrix that transforms A into T_{n-1}. We set $T_0 = A$. The kth step in the iteration is as follows.

Step 1. Denote as A_k the $(n - k + 1) \times (n - k + 1)$ submatrix in the lower right portion of T_{k-1}.
Step 2. Determine an eigenvalue and the corresponding normalized eigenvector for A_k.
Step 3. Construct a unitary matrix N_k which has as its first column the normalized eigenvector found in step 2.
Step 4. For $k = 1$, set $U_1 = N_1$, for $k > 1$, set

$$U_k = \begin{pmatrix} I_{k-1} & 0 \\ 0 & N_k \end{pmatrix}$$

where I_{k-1} is the $(k - 1) \times (k - 1)$ identity matrix.
Step 5. Calculate $T_k = U_k^* T_{k-1} U_k$.

Apply the algorithm to the matrix

$$A = \begin{pmatrix} 1 & 0 & 1 \\ 0 & 1 & 0 \\ 1 & 0 & 1 \end{pmatrix}.$$

Solution 9.   Since $\text{rank}(A) = 2$ one eigenvalue is 0. The corresponding normalized eigenvector of the eigenvalue 0 is

$$\frac{1}{\sqrt{2}}(1 \ \ 0 \ \ -1)^T.$$

Thus we can construct the unitary matrix

$$N_1 = U_1 = \frac{1}{\sqrt{2}} \begin{pmatrix} 1 & 0 & 1 \\ 0 & \sqrt{2} & 0 \\ -1 & 0 & 1 \end{pmatrix}.$$

The two other column vectors in the matrix we construct from the fact that they must be normalized and orthogonal to the first column vector and orthogonal to each other. Now

$$T_1 = U_1^* A U_1 = \begin{pmatrix} 0 & 0 & 0 \\ 0 & 1 & 0 \\ 0 & 0 & 2 \end{pmatrix}$$

which is already the solution of the problem. Thus the eigenvalues of A are $0, 1, 2$.

Problem 10. Let A be an $n \times n$ matrix over **C**. Then there exists an $n \times n$ unitary matrix Q, such that

$$Q^* A Q = D + N$$

where $D = \text{diag}(\lambda_1, \lambda_2, \ldots, \lambda_n)$ is the diagonal matrix composed of the eigenvalues of A and N is a strictly upper triangular matrix (i.e., N has zero entries on the diagonal). The matrix Q is said to provide a *Schur decomposition* of A.

Let

$$A = \begin{pmatrix} 3 & 8 \\ -2 & 3 \end{pmatrix}, \qquad Q = \frac{1}{\sqrt{5}} \begin{pmatrix} 2i & 1 \\ -1 & -2i \end{pmatrix}.$$

Show that Q provides a Schur decomposition of A.

Solution 10. Obviously,

$$Q^* Q = Q Q^* = I_2.$$

Now

$$Q^* A Q = \frac{1}{5} \begin{pmatrix} -2i & -1 \\ 1 & 2i \end{pmatrix} \begin{pmatrix} 3 & 8 \\ -2 & 3 \end{pmatrix} \begin{pmatrix} 2i & 1 \\ -1 & -2i \end{pmatrix}$$

$$= \begin{pmatrix} 3 + 4i & -6 \\ 0 & 3 - 4i \end{pmatrix}$$

$$= \begin{pmatrix} 3 + 4i & 0 \\ 0 & 3 - 4i \end{pmatrix} + \begin{pmatrix} 0 & -6 \\ 0 & 0 \end{pmatrix}.$$

Consequently, we obtained a Schur decomposition of the matrix A with the given Q.

Problem 11. We say that a matrix is upper triangular if all their entries below the main diagonal are 0, and that it is strictly upper triangular if in addition all the entries on the main diagonal are equal to 1. Any invertible real $n \times n$ matrix A can be written as the product of three real $n \times n$ matrices

$$A = ODN$$

where N is strictly upper triangular, D is diagonal with positive entries, and O is orthogonal. This is known as the *Iwasawa decomposition* of the matrix A. The decomposition is unique. In other words, that if $A = O'D'N'$, where O', D' and N' are orthogonal, diagonal with positive entries and strictly upper triangular, respectively, then $O' = O$, $D = D'$ and $N' = N$. Find the Iwasawa decomposition of the matrix

$$A = \begin{pmatrix} 0 & 1 \\ 1 & 2 \end{pmatrix}.$$

Solution 11.   We obtain

$$O = \begin{pmatrix} 0 & 1 \\ 1 & 0 \end{pmatrix}, \quad D = \begin{pmatrix} 1 & 0 \\ 0 & 1 \end{pmatrix}, \quad N = \begin{pmatrix} 1 & 2 \\ 0 & 1 \end{pmatrix}.$$

Problem 12.   Consider the matrix

$$M = \begin{pmatrix} a & b \\ c & d \end{pmatrix}$$

where $a, b, c, d \in \mathbf{C}$ and $ad - bc = 1$. Thus M is an element of the Lie group $SL(2, \mathbf{C})$. The *Iwasawa decomposition* is given by

$$\begin{pmatrix} a & b \\ c & d \end{pmatrix} = \begin{pmatrix} \alpha & \beta \\ -\overline{\beta} & \overline{\alpha} \end{pmatrix} \begin{pmatrix} \delta^{-1/2} & 0 \\ 0 & \delta^{1/2} \end{pmatrix} \begin{pmatrix} 1 & \eta \\ 0 & 1 \end{pmatrix}$$

where $\alpha, \beta, \eta \in \mathbf{C}$ and $\delta \in \mathbf{R}^+$. Find α, β, δ and η.

Solution 12.   Matrix multiplication yields the four equations

$$a = \alpha \delta^{-1/2}$$
$$b = \alpha \eta \delta^{-1/2} + \beta \delta^{1/2}$$
$$c = -\overline{\beta} \delta^{-1/2}$$
$$d = -\overline{\beta} \eta \delta^{-1/2} + \overline{\alpha} \delta^{1/2}$$

where $\alpha \overline{\alpha} + \beta \overline{\beta} = 1$. Solving these four equations yields

$$\delta = (|a|^2 + |b|^2)^{-1}, \quad \alpha = a\delta^{1/2}, \quad \beta = -\overline{c}\delta^{1/2}, \quad \eta = (\overline{a}b + \overline{c}d)\delta.$$

Problem 13.   Let A be a unitary $n \times n$ matrix. Let P be an invertible $n \times n$ matrix. Let $B := AP$. Show that PB^{-1} is unitary.

Solution 13. Since A and P are invertible, the matrix B is also invertible. We have

$$
\begin{aligned}
PB^{-1}(PB^{-1})^* &= PB^{-1}B^{-1*}P^* \\
&= P(AP)^{-1}(AP)^{-1*}P^* \\
&= P(P^{-1}A^{-1})(P^{-1}A^{-1})^*P^* \\
&= P(P^{-1}A^{-1})(A^{-1*}P^{-1*})P^* \\
&= P(P^{-1}A^{-1}AP^{-1*})P^* \\
&= PP^{-1}P^{-1*}P^* \\
&= I_n.
\end{aligned}
$$

Problem 14. Show that every 2×2 matrix A of determinant 1 is the product of three elementary matrices. This means that matrix A can be written as

$$
\begin{pmatrix} a_{11} & a_{12} \\ a_{21} & a_{22} \end{pmatrix} = \begin{pmatrix} 1 & x \\ 0 & 1 \end{pmatrix}\begin{pmatrix} 1 & 0 \\ y & 1 \end{pmatrix}\begin{pmatrix} 1 & z \\ 0 & 1 \end{pmatrix}. \tag{1}
$$

Solution 14. By straightforward computation we find the four equations

$$
a_{11} = 1 + xy, \qquad a_{12} = z(1+xy), \qquad a_{21} = y, \qquad a_{22} = yz + 1. \tag{2}
$$

Case $a_{21} \neq 0$. Then we solve the system of equations (2) for y, x, z in that order using all but the second equation. We obtain

$$
y = a_{21}, \qquad x = \frac{a_{11} - 1}{a_{21}}, \qquad z = \frac{a_{22} - 1}{a_{21}}.
$$

For these choices of x, y, z the second equation of (2) is also satisfied since

$$
z(1+xy)+x = a_{21}^{-1}(a_{22}-1)(1+a_{11}-1)+a_{21}^{-1}(a_{11}-1) = a_{21}^{-1}(a_{11}a_{22}-1) = a_{12}
$$

where we used that $\det A = 1$, i.e. $a_{11}a_{22} - a_{12}a_{21} = 1$.
Case $a_{21} = 0$. We have $y = 0$. Thus we have the representation

$$
\begin{pmatrix} a_{11} & a_{12} \\ 0 & a_{11}^{-1} \end{pmatrix} = \begin{pmatrix} a_{11} & 0 \\ 0 & 1 \end{pmatrix}\begin{pmatrix} 1 & a_{12} \\ 0 & 1 \end{pmatrix}\begin{pmatrix} 1 & 0 \\ 0 & a_{11}^{-1} \end{pmatrix}.
$$

Problem 15. Almost any 2×2 matrix A can be factored (*Gaussian decomposition*) as

$$
\begin{pmatrix} a_{11} & a_{12} \\ a_{21} & a_{22} \end{pmatrix} = \begin{pmatrix} 1 & \alpha \\ 0 & 1 \end{pmatrix}\begin{pmatrix} \lambda & 0 \\ 0 & \mu \end{pmatrix}\begin{pmatrix} 1 & 0 \\ \beta & 1 \end{pmatrix}.
$$

Find the decomposition of the matrix

$$A = \begin{pmatrix} 1 & 1 \\ 1 & 1 \end{pmatrix}.$$

Solution 15.   Matrix multiplication yields

$$\begin{pmatrix} 1 & 1 \\ 1 & 1 \end{pmatrix} = \begin{pmatrix} \lambda + \mu\alpha\beta & \alpha\mu \\ \beta\mu & \mu \end{pmatrix}.$$

Thus we obtain the four equations

$$\lambda + \alpha\beta\mu = 1, \quad \alpha\mu = 1, \quad \beta\mu = 1, \quad \mu = 1.$$

This leads to the unique solution

$$\mu = \alpha = \beta = 1, \qquad \lambda = 0.$$

Thus we have the decomposition

$$\begin{pmatrix} 1 & 1 \\ 1 & 1 \end{pmatrix} = \begin{pmatrix} 1 & 1 \\ 0 & 1 \end{pmatrix} \begin{pmatrix} 0 & 0 \\ 0 & 1 \end{pmatrix} \begin{pmatrix} 1 & 0 \\ 1 & 1 \end{pmatrix}.$$

Problem 16.   Let A be an $n \times n$ matrix over **R**. Consider the *LU*-decomposition $A = LU$, where L is a unit lower triangular matrix and U is an upper triangular matrix. The *LDU*-decomposition is defined as $A = LDU$, where L is unit lower triangular, D is diagonal and U is unit upper triangular. Let

$$A = \begin{pmatrix} 2 & 4 & -2 \\ 4 & 9 & -3 \\ -2 & -3 & 7 \end{pmatrix}.$$

Find the *LDU*-decomposition via the *LU*-decomposition.

Solution 16.   We have the *LU*-decomposition

$$L = \begin{pmatrix} 1 & 0 & 0 \\ 2 & 1 & 0 \\ -1 & 1 & 1 \end{pmatrix}, \qquad U = \begin{pmatrix} 2 & 4 & -2 \\ 0 & 1 & 1 \\ 0 & 0 & 4 \end{pmatrix}.$$

Then the *LDU*-decomposition follows as

$$A = \begin{pmatrix} 1 & 0 & 0 \\ 2 & 1 & 0 \\ -1 & 1 & 1 \end{pmatrix} \begin{pmatrix} 2 & 0 & 0 \\ 0 & 1 & 0 \\ 0 & 0 & 4 \end{pmatrix} \begin{pmatrix} 1 & 2 & -1 \\ 0 & 1 & 1 \\ 0 & 0 & 1 \end{pmatrix}.$$

Problem 17. Let U be an $n \times n$ unitary matrix. The matrix U can always be diagonalized by a unitary matrix V such that

$$U = V \begin{pmatrix} e^{i\theta_1} & \cdots & 0 \\ \vdots & \ddots & \vdots \\ 0 & \cdots & e^{i\theta_n} \end{pmatrix} V^*$$

where $e^{i\theta_j}$, $\theta_j \in [0, 2\pi)$ are the eigenvalues of U. Let

$$U = \begin{pmatrix} 0 & 1 \\ 1 & 0 \end{pmatrix}.$$

Thus the eigenvalues are 1 and -1. Find the unitary matrix V such that

$$U = \begin{pmatrix} 0 & 1 \\ 1 & 0 \end{pmatrix} = V \begin{pmatrix} 1 & 0 \\ 0 & -1 \end{pmatrix} V^*.$$

Solution 17. We find the Hadamard matrix

$$V = V^* = \frac{1}{\sqrt{2}} \begin{pmatrix} 1 & 1 \\ 1 & -1 \end{pmatrix}.$$

Chapter 7

Functions of Matrices

Let p be a polynomial

$$p(x) = \sum_{j=0}^{n} c_j x^j$$

then the corresponding matrix function of an $n \times n$ matrix A can be defined by

$$p(A) = \sum_{j=0}^{n} c_j A^j$$

with the convention $A^0 = I_n$. If a function f of a complex variable z has a *MacLaurin series expansion*

$$f(z) = \sum_{j=0}^{\infty} c_j z^j$$

which converges for $|z| < R$ ($R > 0$), then the matrix series

$$f(A) = \sum_{j=0}^{\infty} c_j A^j$$

converges, provided A is square and each of its eigenvalues has absolute value less than R. The most used matrix function is $\exp(A)$ defined by

$$\exp(A) := \sum_{j=0}^{\infty} \frac{A^j}{j!} \equiv \lim_{n \to \infty} \left(I_n + \frac{A}{n} \right)^n .$$

Problem 1. Let

$$A = \begin{pmatrix} 1 & 1 \\ 0 & 1 \end{pmatrix}.$$

Calculate $f(A) = A^3$ using the *Cayley-Hamilton theorem*. The Cayley-Hamilton theorem states that substituting the matrix in the characteristic polynomial results in the zero matrix.

Solution 1. The characteristic equation of A

$$\det(A - \lambda I_2) = 0$$

is given by

$$\lambda^2 - 2\lambda + 1 = 0.$$

Thus from Cayley-Hamilton theorem we obtain

$$A^2 = 2A - I_2.$$

Using this expression it follows that

$$A^3 = AA^2 = A(2A - I_2) = 2A^2 - A = 3A - 2I_2.$$

Consequently

$$A^3 = \begin{pmatrix} 3 & 3 \\ 0 & 3 \end{pmatrix} - \begin{pmatrix} 2 & 0 \\ 0 & 2 \end{pmatrix} = \begin{pmatrix} 1 & 3 \\ 0 & 1 \end{pmatrix}.$$

Problem 2. Let A be an $n \times n$ matrix over $\mathbf{C}$ with $A^2 = rA$, where $r \in \mathbf{C}$ and $r \neq 0$.
(i) Calculate e^{zA}, where $z \in \mathbf{C}$.
(ii) Let $U(z) = e^{zA}$. Let $z' \in \mathbf{C}$. Calculate $U(z)U(z')$.

Solution 2. (i) From $A^2 = rA$, we obtain

$$A^3 = r^2 A, \quad A^4 = r^3 A, \quad \cdots \quad , A^n = r^{n-1} A.$$

Thus we obtain

$$e^{zA} = \sum_{j=0}^{\infty} \frac{(zA)^j}{j!}$$

$$= I_n + \frac{z}{1!}A + \frac{rz^2}{2!}A + \frac{r^2 z^3}{3!}A + \cdots + \frac{r^{n-1} z^n}{n!}A + \cdots$$

$$= I_n + A\left(\frac{z}{1!} + \frac{rz^2}{2!} + \frac{r^2 z^3}{3!} + \cdots + \frac{r^{n-1} z^n}{n!} + \cdots \right)$$

$$= I_n + \frac{A}{r}\left(rz + \frac{(rz)^2}{2!} + \frac{(rz)^3}{3!} + \cdots + \frac{(rz)^n}{n!} + \cdots\right)$$

$$= I_n + \frac{A}{r}\left(1 + rz + \frac{(rz)^2}{2!} + \frac{(rz)^3}{3!} + \cdots + \frac{(rz)^n}{n!} + \cdots\right) - \frac{A}{r}$$

$$= I_n + \frac{A}{r}e^{rz} - \frac{A}{r}$$

$$= I_n + \frac{A}{r}(e^{rz} - 1).$$

(ii) Straightforward calculation yields

$$U(z)U(z') = I_n + \frac{A}{r}(e^{r(z+z')} - 1) = U(z + z').$$

This result is also obvious since $[A, A] = 0_n$, where $[\,,\,]$ denotes the commutator and therefore

$$e^{zA}e^{z'A} = e^{(z+z')A}.$$

Problem 3.   Let A be an $n \times n$ matrix over **C**. We define $\sin(A)$ as

$$\sin(A) := \sum_{j=0}^{\infty} \frac{(-1)^j}{(2j+1)!} A^{2j+1}.$$

Can we find a 2×2 matrix B over the real numbers **R** such that

$$\sin(B) = \begin{pmatrix} 1 & 4 \\ 0 & 1 \end{pmatrix}? \tag{1}$$

Solution 3.   Owing to the right-hand side of equation (1) the 2×2 matrix B can only be of the form

$$B = \begin{pmatrix} x & y \\ 0 & x \end{pmatrix}$$

with $x \neq 0$. Straightforward calculation yields

$$\sin\begin{pmatrix} x & y \\ 0 & x \end{pmatrix} = \sum_{k=0}^{\infty} \frac{(-1)^k}{(2k+1)!} \begin{pmatrix} x & y \\ 0 & x \end{pmatrix}^{2k+1}$$

$$= \sum_{k=0}^{\infty} \frac{(-1)^k x^{2k+1}}{(2k+1)!} \begin{pmatrix} 1 & y/x \\ 0 & 1 \end{pmatrix}^{2k+1}$$

$$= \sum_{k=0}^{\infty} \frac{(-1)^k x^{2k+1}}{(2k+1)!} \begin{pmatrix} 1 & (2k+1)y/x \\ 0 & 1 \end{pmatrix}$$

$$= \begin{pmatrix} \sin x & y \cos x \\ 0 & \sin x \end{pmatrix}.$$

Thus $\sin x = 1$ and $y \cos x = 4$. From the first condition we find that $\cos x = 0$ and therefore the second condition $y \cos x = 4$ cannot be satisfied. Thus there is no such matrix.

Problem 4. Let A be an $n \times n$ matrix over $\mathbf{C}$ and

$$p(\lambda) = \det(A - \lambda I_n).$$

Show that $p(A) = 0_n$, i.e., the matrix A satisfies its *characteristic equation* (*Cayley-Hamilton theorem*).

Solution 4. There exists an invertible $n \times n$ matrix X over $\mathbf{C}$ such that

$$X^{-1}AX = J = \mathrm{diag}(J_1, J_2, \ldots, J_t)$$

where

$$J_i = \begin{pmatrix} \lambda_i & 1 & 0 & \cdots & 0 \\ 0 & \lambda_i & 1 & \ddots & \vdots \\ \vdots & \ddots & \ddots & \ddots & \vdots \\ \vdots & & \ddots & \ddots & 1 \\ 0 & \cdots & \cdots & 0 & \lambda_i \end{pmatrix}$$

is an upper bidiagonal $m_i \times m_i$ matrix (*Jordan block*) that has on its main diagonal the eigenvalue λ_i of the matrix A (at least m_i-multiple eigenvalue of the matrix A since to this eigenvalue may correspond some more Jordan blocks) and $m_1 + \cdots + m_t = n$. Since $J_i - \lambda I_n = (\delta_{k,j-1})$, then $(\delta_{k,j-1})^{m_i} = 0$ and $(J_i - \lambda_i I_n)^{m_i} = 0$. If $p(\lambda)$ is the *characteristic polynomial* of the matrix A and the zeros of this polynomial are $\lambda_1, \ldots, \lambda_t$, then

$$p(\lambda) = (-1)^n (\lambda - \lambda_1)^{m_1} (\lambda - \lambda_2)^{m_2} \cdots (\lambda - \lambda_t)^{m_t}$$

and

$$p(J) = (-1)^n (J - \lambda_1 I_n)^{m_1} (J - \lambda_2 I_n)^{m_2} \cdots (J - \lambda_t I_n)^{m_t}.$$

We show that $p(J) = 0$. Let the matrix J have the block form

$$J = \begin{pmatrix} J_1 & 0 & 0 & \cdots & 0 \\ 0 & J_2 & 0 & \cdots & 0 \\ 0 & 0 & J_3 & \cdots & 0 \\ \vdots & \vdots & \vdots & \ddots & \vdots \\ 0 & 0 & 0 & \cdots & J_t \end{pmatrix}.$$

Then we obtain

$$p(J) = (-1)^n (J - \lambda_1 I_n)^{m_1} (J - \lambda_2 I_n)^{m_2} \cdots (J - \lambda_t I_n)^{m_t}$$

$$= (-1)^n \prod_{k=1}^{t} (J - \lambda_k I_n)^{m_k}$$

$$= (-1)^n \prod_{k=1}^{t} \begin{pmatrix} J_1 - \lambda_k I_{m_k} & 0 & \cdots & 0 \\ 0 & J_2 - \lambda_k I_{m_k} & \cdots & 0 \\ \vdots & \vdots & \ddots & \vdots \\ 0 & 0 & \cdots & J_t - \lambda_k I_{m_k} \end{pmatrix}^{m_k}$$

$$= (-1)^n \prod_{k=1}^{t} \begin{pmatrix} (J_1 - \lambda_k I_{m_k})^{m_k} & 0 & \cdots & 0 \\ 0 & (J_2 - \lambda_k I_{m_k})^{m_k} & \cdots & 0 \\ \vdots & \vdots & \ddots & \vdots \\ 0 & 0 & \cdots & (J_t - \lambda_k I_{m_k})^{m_k} \end{pmatrix}$$

$$= \begin{pmatrix} 0 & 0 & \cdots & 0 \\ 0 & 0 & \cdots & 0 \\ \vdots & \vdots & \ddots & \vdots \\ 0 & 0 & \cdots & 0 \end{pmatrix}.$$

From the $X^{-1}AX = J$ it follows that $A = XJX^{-1}$. Thus

$$\begin{aligned} p(A) &= p(XJX^{-1}) \\ &= (-1)^n(XJX^{-1} - X\lambda_1 I_n X^{-1})(XJX^{-1} - X\lambda_2 I_n X^{-1}) \\ &\quad \cdots (XJX^{-1} - X\lambda_n I_n X^{-1}) \\ &= (-1)^n X(J - \lambda_1 I_n)X^{-1} X(J - \lambda_2 I_n)X^{-1} \cdots X(J - \lambda_n I_n)X^{-1} \\ &= Xp(J)X^{-1} \\ &= 0_n \end{aligned}$$

since $p(J) = 0_n$.

Problem 5. Consider the unitary matrix

$$U = \begin{pmatrix} 0 & 1 \\ 1 & 0 \end{pmatrix}.$$

Can we find an $\alpha \in \mathbf{R}$ such that $U = \exp(\alpha A)$, where

$$A = \begin{pmatrix} 0 & 1 \\ -1 & 0 \end{pmatrix}?$$

Solution 5. Since $A^2 = -I_2$, $A^3 = -A$, $A^4 = I_2$ etc. we obtain

$$e^{\alpha A} = I_2 \cos\alpha + A \sin\alpha.$$

This leads to the matrix equation

$$\begin{pmatrix} \cos\alpha & \sin\alpha \\ -\sin\alpha & \cos\alpha \end{pmatrix} = \begin{pmatrix} 0 & 1 \\ 1 & 0 \end{pmatrix}$$

which has no solution.

Problem 6. Let A be an $n \times n$ matrix over $\mathbf{C}$. Assume that $A^2 = cI_n$, where $c \in \mathbf{R}$.
(i) Calculate $\exp(A)$.
(ii) Apply the result to the 2×2 matrix ($z \neq 0$)

$$B = \begin{pmatrix} 0 & z \\ -\bar{z} & 0 \end{pmatrix}.$$

Thus B is skew-hermitian, i.e., $\bar{B}^T = -B$.

Solution 6. (i) We have $A^3 = cA$, $A^4 = c^2 I_n$, $A^5 = c^2 A$, $A^6 = c^3 I_n$. Thus, in general

$$A^m = \begin{cases} c^{m/2} I_n & m \text{ even} \\ c^{(m-1)/2} A & m \text{ odd}. \end{cases}$$

It follows that

$$e^A = I_n + A + \frac{c}{2!}I_n + \frac{c}{3!}A + \frac{c^2}{4!}I_n + \frac{c^2}{5!}A + \frac{c^3}{6!}I_n \, |$$

$$= I_n \left(1 + \frac{c}{2!} + \frac{c^2}{4!} + \frac{c^3}{6!} + \frac{c^4}{8!} + \cdots \right)$$

$$+ A \left(1 + \frac{c}{3!} + \frac{c^2}{5!} + \frac{c^3}{7!} + \frac{c^4}{9!} + \cdots \right)$$

$$= I_n \left(1 + \frac{c}{2!} + \frac{c^2}{4!} + \frac{c^3}{6!} + \frac{c^4}{8!} + \cdots \right)$$

$$+ \frac{A}{\sqrt{c}} \left(\sqrt{c} + \frac{c\sqrt{c}}{3!} + \frac{c^2\sqrt{c}}{5!} + \frac{c^3\sqrt{c}}{7!} + \frac{c^4\sqrt{c}}{9!} + \cdots \right)$$

$$= I_n \cosh(\sqrt{c}) + \frac{A}{\sqrt{c}} \sinh(\sqrt{c}).$$

(ii) We have

$$B^2 = -rI_2, \qquad r = z^*z, \qquad r > 0.$$

Thus the condition $B^2 = cI_2$ is satisfied with $c = -r$. It follows that

$$e^B = I_2 \cosh(\sqrt{-r}) + \frac{B}{\sqrt{-r}} \sinh(\sqrt{-r}).$$

With $\sqrt{-r} = i\sqrt{r}$, $\cosh(i\sqrt{r}) = \cos(\sqrt{r})$, $\sinh(i\sqrt{r}) = i\sin(\sqrt{r})$ we obtain

$$e^B = I_2 \cos(\sqrt{r}) + \frac{B}{\sqrt{r}} \sin(\sqrt{r}).$$

Problem 7.   Let H be a hermitian matrix, i.e., $H = H^*$. It is known that $U := e^{iH}$ is a unitary matrix. Let

$$H = \begin{pmatrix} a & b \\ \bar{b} & a \end{pmatrix}, \qquad a \in \mathbf{R}, \ b \in \mathbf{C}$$

with $b \neq 0$.

(i) Calculate e^{iH} using the normalized eigenvectors of H to construct a unitary matrix V such that $V^* HV$ is a diagonal matrix.

(ii) Specify a, b such that we find the unitary matrix

$$U = \begin{pmatrix} 0 & 1 \\ 1 & 0 \end{pmatrix}.$$

Solution 7.   (i) The eigenvalues of H are given by

$$\lambda_1 = a + \sqrt{b\bar{b}}, \qquad \lambda_2 = a - \sqrt{b\bar{b}}.$$

The corresponding normalized eigenvectors are

$$\mathbf{x}_1 = \frac{1}{\sqrt{2}} \begin{pmatrix} 1 \\ \sqrt{b\bar{b}}/b \end{pmatrix}, \qquad \mathbf{x}_2 = \frac{1}{\sqrt{2}} \begin{pmatrix} 1 \\ -\sqrt{b\bar{b}}/b \end{pmatrix}.$$

Thus the unitary matrices V, V^* which diagonalize H are

$$V = \frac{1}{\sqrt{2}} \begin{pmatrix} 1 & 1 \\ \sqrt{b\bar{b}}/b & -\sqrt{b\bar{b}}/b \end{pmatrix}, \qquad V^* = \frac{1}{\sqrt{2}} \begin{pmatrix} 1 & b/\sqrt{b\bar{b}} \\ 1 & -b/\sqrt{b\bar{b}} \end{pmatrix}$$

with

$$D := V^* HV = \begin{pmatrix} a + \sqrt{b\bar{b}} & 0 \\ 0 & a - \sqrt{b\bar{b}} \end{pmatrix}.$$

From $U = e^{iH}$ it follows that

$$V^* UV = V^* e^{iH} V = e^{iV^* HV} = e^{iD}.$$

Thus

$$e^{iD} = \begin{pmatrix} e^{i(a+\sqrt{b\bar{b}})} & 0 \\ 0 & e^{i(a-\sqrt{b\bar{b}})} \end{pmatrix}$$

and since $V^* = V^{-1}$ the unitary matrix U is given by

$$U = Ve^{iD}V^*.$$

We obtain

$$U = e^{ia} \begin{pmatrix} \cos(\sqrt{b\bar{b}}) & b/\sqrt{b\bar{b}}\sin(\sqrt{b\bar{b}}) \\ \sqrt{b\bar{b}}/b\sin(\sqrt{b\bar{b}}) & \cos(\sqrt{b\bar{b}}) \end{pmatrix}.$$

(ii) If $a = \pi/2$ and $b = -\pi/2$ we find the unitary matrix, where we used that $\cos(\pi/2) = 0$.

Problem 8. It is known that any $n \times n$ unitary matrix U can be written as $U = \exp(iK)$, where K is a hermitian matrix. Assume that $\det U = -1$. What can be said about the trace of K?

Solution 8. For any $n \times n$ matrix A we have the identity

$$\det e^A \equiv e^{\operatorname{tr} A}$$

where tr denotes the trace of the matrix. Thus it follows that

$$-1 = \det U = \det e^{iK} = e^{i\operatorname{tr} K}.$$

Thus $\operatorname{tr} K = \pi \ (\pm n 2\pi)$ with $n \in \mathbf{N}$.

Problem 9. The *MacLaurin series* for $\arctan(z)$ is defined as

$$\arctan(z) = z - \frac{z^3}{3} + \frac{z^5}{5} - \frac{z^7}{7} + \cdots = \sum_{j=0}^{\infty} \frac{(-1)^j z^{2j+1}}{2j+1}$$

which converges for all complex values of z having absolute value less than 1, i.e., $|z| < 1$. Let A be an $n \times n$ matrix. Thus the series expansion

$$\arctan(A) = A - \frac{A^3}{3} + \frac{A^5}{5} - \frac{A^7}{7} + \cdots = \sum_{j=0}^{\infty} \frac{(-1)^j A^{2j+1}}{2j+1}$$

is well-defined for A if all eigenvalues λ of A satisfy $|\lambda| < 1$. Let

$$A = \frac{1}{2} \begin{pmatrix} 1 & 1 \\ 1 & 1 \end{pmatrix}.$$

Does $\arctan(A)$ exist?

Solution 9. The eigenvalues of A are 0 and 1. Thus $\arctan(A)$ does not exist.

Problem 10. Let A be an $n \times n$ positive definite matrix over $\mathbf{R}$, i.e. $\mathbf{x}^T A \mathbf{x} > 0$ for all $\mathbf{x} \in \mathbf{R}^n$. Calculate

$$\int_{\mathbf{R}^n} \exp(-\mathbf{x}^T A \mathbf{x}) dx.$$

Solution 10. Since A is positive definite it can be written in the form $A = S^T S$, where T denotes transpose and $S \in GL(n, \mathbf{R})$. Thus

$$|\det S| = \sqrt{\det(A)}.$$

Let $\mathbf{y} = S\mathbf{x}$. Now

$$\int_{\mathbf{R}^n} \exp(-\mathbf{x}^T A\mathbf{x})d\mathbf{x} = \int_{\mathbf{R}^n} \exp(-(S\mathbf{x})^T(S\mathbf{x}))d\mathbf{x}$$

$$= \int_{\mathbf{R}^n} \exp(-\mathbf{y}^T\mathbf{y})d(S^{-1}\mathbf{y})$$

$$= |\det S^{-1}| \int_{\mathbf{R}^n} \exp(-\mathbf{y}^T\mathbf{y})d\mathbf{y}$$

$$= \frac{\pi^{n/2}}{\sqrt{\det(A)}}.$$

Problem 11. For every positive definite matrix A, there is a unique positive definite matrix Q such that $Q^2 = A$. The matrix Q is called the *square root* of A. Can we find the square root of the matrix

$$B = \frac{1}{2}\begin{pmatrix} 5 & 3 \\ 3 & 5 \end{pmatrix} ?$$

Solution 11. First we see that B is symmetric over $\mathbf{R}$ and therefore hermitian. The matrix is also positive definite since the eigenvalues are 1 and 4. Thus we can find an orthogonal matrix U such that $U^T BU = \mathrm{diag}(4, 1)$ with U given by

$$U = \frac{1}{\sqrt{2}}\begin{pmatrix} 1 & -1 \\ 1 & 1 \end{pmatrix}.$$

Obviously the square root of the diagonal matrix $\mathrm{diag}(4, 1)$ is $D = \mathrm{diag}(2, 1)$. Then the square root of B is

$$Q = UDU^T = \frac{1}{2}\begin{pmatrix} 3 & 1 \\ 1 & 3 \end{pmatrix}.$$

Problem 12. Let A, B be $n \times n$ matrices over $\mathbf{C}$. Assume that

$$[A, [A, B]] = [B, [A, B]] = 0_n. \tag{1}$$

Show that

$$e^{A+B} = e^A e^B e^{-\frac{1}{2}[A,B]} \tag{2a}$$

$$e^{A+B} = e^B e^A e^{+\frac{1}{2}[A,B]} .$$ (2b)

Use the *technique of parameter differentiation*, i.e. consider the matrix-valued function

$$f(\epsilon) := e^{\epsilon A} e^{\epsilon B}$$

where ϵ is a real parameter. Then take the derivative of f with respect to ϵ.

Solution 12. If we differentiate f with respect to ϵ we find

$$\frac{df}{d\epsilon} = A e^{\epsilon A} e^{\epsilon B} + e^{\epsilon A} e^{\epsilon B} B = (A + e^{\epsilon A} B e^{-\epsilon A}) f(\epsilon)$$

since $e^{\epsilon A} e^{-\epsilon A} = I_n$. We have

$$e^{\epsilon A} B \epsilon^{-\epsilon A} = B + \epsilon[A, B]$$

where we have taken (1) into account. Thus we obtain the linear matrix-valued differential equation

$$\frac{df}{d\epsilon} = ((A + B) + \epsilon[A, B]) f(\epsilon) .$$

Since the matrix $A + B$ commutes with $[A, B]$ we may treat $A + B$ and $[A, B]$ as ordinary commuting variables and integrate this linear differential equation with the initial condition $f(0) = I_n$. We find

$$f(\epsilon) = e^{\epsilon(A+B) + (\epsilon^2/2)[A,B]} = e^{\epsilon(A+B)} e^{(\epsilon^2/2)[A,B]}$$

where the last form follows since $A + B$ commutes with $[A, B]$. If we set $\epsilon = 1$ and multiply both sides by $e^{-[A,B]/2}$ then (2a) follows. Likewise we can prove the second form of the identity (2b).

Problem 13. Let

$$J^+ := \begin{pmatrix} 0 & 1 \\ 0 & 0 \end{pmatrix}, \qquad J^- := \begin{pmatrix} 0 & 0 \\ 1 & 0 \end{pmatrix}, \qquad J_3 := \frac{1}{2} \begin{pmatrix} 1 & 0 \\ 0 & -1 \end{pmatrix} .$$

(i) Let $\epsilon \in \mathbf{R}$. Find

$$e^{\epsilon J^+}, \quad e^{\epsilon J^-}, \quad e^{\epsilon(J^+ + J^-)} .$$

(ii) Let $r \in \mathbf{R}$. Show that

$$e^{r(J^+ + J^-)} \equiv e^{J^- \tanh(r)} e^{2J_3 \ln(\cosh(r))} e^{J^+ \tanh(r)} .$$

Solution 13. (i) Using the expansion for an $n \times n$ matrix A

$$\exp(\epsilon A) = \sum_{j=0}^{\infty} \frac{\epsilon^j A^j}{j!}$$

we find

$$e^{\epsilon J^+} = \begin{pmatrix} 1 & 0 \\ 0 & 1 \end{pmatrix} + \epsilon \begin{pmatrix} 0 & 1 \\ 0 & 0 \end{pmatrix}, \qquad e^{\epsilon J^-} = \begin{pmatrix} 1 & 0 \\ 0 & 1 \end{pmatrix} + \epsilon \begin{pmatrix} 0 & 0 \\ 1 & 0 \end{pmatrix}$$

and

$$e^{\epsilon(J^+ + J^-)} = \begin{pmatrix} 1 & 0 \\ 0 & 1 \end{pmatrix} \cosh(\epsilon) + \begin{pmatrix} 0 & 1 \\ 1 & 0 \end{pmatrix} \sinh(\epsilon).$$

(ii) Using the results from (i) we find the identity.

Problem 14.   Let A, B, C_2, ... , C_m, ... be $n \times n$ matrices over **C**. The *Zassenhaus formula* is given by

$$\exp(A + B) = \exp(A) \exp(B) \exp(C_2) \cdots \exp(C_m) \cdots$$

The left-hand side is called the *disentangled form* and the right-hand side is called the *undisentangled form*. Find C_2, C_3, ..., using the *comparison method*. In the comparison method the disentangled and undisentangled forms are expanded in terms of an ordering scalar α and matrix coefficients of equal powers of α are compared. From

$$\exp(\alpha(A + B)) = \exp(\alpha A) \exp(\alpha B) \exp(\alpha^2 C_2) \exp(\alpha^3 C_3) \cdots$$

we obtain

$$\sum_{k=0}^{\infty} \frac{\alpha^k}{k!} (A + B)^k = \sum_{r_0, r_1, r_2, r_3, \ldots = 0}^{\infty} \frac{\alpha^{r_0 + r_1 + 2r_2 + 3r_3 + \ldots}}{r_0! r_1! r_2! r_3! \cdots} A^{r_0} B^{r_1} C_2^{r_2} C_3^{r_3} \cdots$$

(i) Find C_2 and C_3.
(ii) Assume that $[A, [A, B]] = 0_n$ and $[B, [A, B]] = 0_n$. What conclusion can we draw for the Zassenhaus formula?

Solution 14.   (i) For α^2 we have the decompositions $(r_0, r_1, r_2) = (2, 0, 0)$, $(1, 1, 0)$, $(0, 2, 0)$, $(0, 0, 1)$. Thus we obtain

$$(A + B)^2 = A^2 + 2AB + B^2 + 2C_2.$$

Thus it follows that

$$C_2 = -\frac{1}{2}[A, B].$$

For α^3 we obtain

$$(A + B)^3 = A^3 + 3A^2 B + 3AB^2 + B^3 + 6AC_2 + 6BC_2 + 6C_3.$$

Using C_2 given above we obtain

$$C_3 = \frac{1}{3}[B, [A, B]] + \frac{1}{6}[A, [A, B]].$$

(ii) Since $[B, [A, B]] = 0_n$, $[A, [A, B]] = 0_n$ we find that $C_3 = C_4 = \cdots = 0_n$. Thus

$$\exp(\alpha(A + B)) = \exp(\alpha A) \exp(\alpha B) \exp(-\alpha^2 [A, B]/2) \,.$$

Problem 15. Calculating $\exp(A)$ we can also use the Cayley-Hamilton theorem and the *Putzer method*. The Putzer method is as follows. Using the Cayley-Hamilton theorem we can write

$$f(A) = a_{n-1} A^{n-1} + a_{n-2} A^{n-2} + \cdots + a_2 A^2 + a_1 A + a_0 I_n \qquad (1)$$

where the complex numbers $a_0, a_1, \ldots, a_{n-1}$ are determined as follows: Let

$$r(\lambda) := a_{n-1} \lambda^{n-1} + a_{n-2} \lambda^{n-2} + \cdots + a_2 \lambda^2 + a_1 \lambda + a_0$$

which is the right-hand side of (1) with A^j replaced by λ^j ($j = 0, 1, \ldots, n - 1$). For each distinct eigenvalue λ_j of the matrix A, we consider the equation

$$f(\lambda_j) = r(\lambda_j) \,. \qquad (2)$$

If λ_j is an eigenvalue of multiplicity k, for $k > 1$, then we consider also the following equations

$$f'(\lambda)|_{\lambda=\lambda_j} = r'(\lambda)|_{\lambda=\lambda_j}, \quad \cdots \quad , f^{(k-1)}(\lambda)\Big|_{\lambda=\lambda_j} = r^{(k-1)}(\lambda)\Big|_{\lambda=\lambda_j} \,.$$

Calculate $\exp(A)$ with

$$A = \begin{pmatrix} 2 & -1 \\ -1 & 2 \end{pmatrix}$$

with the Putzer method.

Solution 15. The eigenvalues of A are given by $\lambda_1 = 3$ and $\lambda_2 = 1$. Since we have a 2×2 matrix we can write

$$e^A = c_1 A + c_0 I_2 = \begin{pmatrix} 2c_1 + c_0 & -c_1 \\ -c_1 & 2c_1 + c_0 \end{pmatrix} \,.$$

The coefficients c_1 and c_0 are determined by

$$r(\lambda_1) = c_1 \lambda_1 + c_0 \Rightarrow e^3 = 3c_1 + c_0$$

$$r(\lambda_2) = c_1 \lambda_2 + c_0 \Rightarrow e = c_1 + c_0 \,.$$

The solution of this system of linear equations is

$$c_1 = \frac{1}{2} \left(e^3 - e \right), \qquad c_0 = -\frac{1}{2} e^3 + \frac{3}{2} e \,.$$

Thus

$$e^A = \frac{1}{2}e \begin{pmatrix} e^2 + 1 & 1 - e^2 \\ 1 - e^2 & e^2 + 1 \end{pmatrix}.$$

Problem 16. Any unitary matrix U can be written as $U = \exp(iK)$, where K is hermitian. Apply the method of the previous problem to find K for the Hadamard matrix

$$U_H = \frac{1}{\sqrt{2}} \begin{pmatrix} 1 & 1 \\ 1 & -1 \end{pmatrix}.$$

Solution 16. The hermitian 2×2 matrix K is given by

$$K = \begin{pmatrix} a & b \\ \bar{b} & c \end{pmatrix}, \qquad a, c \in \mathbf{R}, \quad b \in \mathbf{C}.$$

Then we find the condition on a, b and c such that $e^{iK} = U_H$. The eigenvalues of iK are given by

$$\lambda_{1,2} = \frac{i(a+c)}{2} \pm \frac{1}{2}\sqrt{2ac - a^2 - c^2 - 4b\bar{b}}.$$

We set in the following

$$\Delta := \lambda_1 - \lambda_2 = \sqrt{2ac - a^2 - c^2 - 4b\bar{b}}.$$

To apply the method given above we have

$$r(\lambda) = \alpha_1 \lambda + \alpha_0 = f(\lambda) = e^\lambda.$$

Thus we obtain the two equations

$$e^{\lambda_1} = \alpha_1 \lambda_1 + \alpha_0, \qquad e^{\lambda_2} = \alpha_1 \lambda_2 + \alpha_0.$$

It follows that

$$\alpha_1 = \frac{e^{\lambda_1} - e^{\lambda_2}}{\lambda_1 - \lambda_2}, \qquad \alpha_0 = \frac{e^{\lambda_2}\lambda_1 - e^{\lambda_1}\lambda_2}{\lambda_1 - \lambda_2}.$$

Thus we have the condition

$$e^{iK} = \alpha_1 iK + \alpha_0 I_2 = \begin{pmatrix} i\alpha_1 a + \alpha_0 & i\alpha_1 b \\ i\alpha_1 \bar{b} & i\alpha_1 c + \alpha_0 \end{pmatrix} = \frac{1}{\sqrt{2}} \begin{pmatrix} 1 & 1 \\ 1 & -1 \end{pmatrix}.$$

We obtain the four equations

$$i\alpha_1 a + \alpha_0 = \frac{1}{\sqrt{2}}, \quad i\alpha_1 c + \alpha_0 = -\frac{1}{\sqrt{2}}, \quad i\alpha_1 b = \frac{1}{\sqrt{2}}, \quad i\alpha_1 \bar{b} = \frac{1}{\sqrt{2}}.$$

From the last two equations we find that $\bar{b} = b$, i.e., b is real. From the first two equations we find $\alpha_0 = -i\alpha_1(a+c)/2$ and therefore, using the last two equations, $c = a - 2b$. Thus

$$\begin{pmatrix} i\alpha_1 a + \alpha_0 & i\alpha_1 b \\ i\alpha_1 \bar{b} & i\alpha_1 c + \alpha_0 \end{pmatrix} = \begin{pmatrix} i\alpha_1 b & i\alpha_1 b \\ i\alpha_1 b & -i\alpha_1 b \end{pmatrix}.$$

From the eigenvalues of e^{iK} we find $e^{\lambda_1} - e^{\lambda_2} = 2$ and

$$\Delta = \sqrt{2ac - a^2 - c^2 - 4b^2} = 2\sqrt{2}ib.$$

Furthermore

$$\lambda_1 = i(a-b) + \sqrt{2}ib, \qquad \lambda_2 = i(a-b) - \sqrt{2}ib.$$

Thus we arrive at the equation

$$e^{i(a-b)+\sqrt{2}ib} - e^{i(a-b)-\sqrt{2}ib} = 2.$$

It follows that

$$ie^{i(a-b)} \sin(\sqrt{2}b) = 1$$

and therefore

$$i\cos(a-b)\sin(\sqrt{2}b) - \sin(a-b)\sin(\sqrt{2}b) = 1$$

with a solution

$$b = \frac{\pi}{2\sqrt{2}}, \qquad a = \frac{\pi}{2}\left(3 + \frac{1}{\sqrt{2}}\right), \qquad c = a - 2b = \frac{\pi}{2}\left(3 - \frac{1}{\sqrt{2}}\right).$$

Then the matrix K is given by

$$K = \frac{\pi}{2}\begin{pmatrix} 3 + 1/\sqrt{2} & 1/\sqrt{2} \\ 1/\sqrt{2} & 3 - 1/\sqrt{2} \end{pmatrix} = \frac{3\pi}{2}\begin{pmatrix} 1 & 0 \\ 0 & 1 \end{pmatrix} + \frac{\pi}{2} \cdot \frac{1}{\sqrt{2}}\begin{pmatrix} 1 & 1 \\ 1 & -1 \end{pmatrix}.$$

We note that the second matrix on the right-hand side is the Hadamard matrix again.

Problem 17. Let A, B be $n \times n$ matrices and $t \in \mathbf{R}$. Show that

$$e^{t(A+B)} - e^{tA}e^{tB} = \frac{t^2}{2}(BA - AB) + \text{higher order terms in } t. \qquad (1)$$

Solution 17. We have

$$e^{tA}e^{tB} = \left(I_n + tA + \frac{t^2}{2!}A^2 + \cdots\right)\left(I_n + tB + \frac{t^2}{2!}B^2 + \cdots\right)$$

$$= I_n + t(A + B) + \frac{t^2}{2!}(A^2 + B^2 + 2AB) + \cdots$$

and

$$e^{t(A+B)} = I_n + t(A + B) + \frac{t^2}{2!}(A + B)^2 + \cdots$$

$$= I_n + t(A + B) + \frac{t^2}{2!}(A^2 + B^2 + AB + BA) + \cdots .$$

Subtracting the two equations yields (1).

Problem 18.   Let K be an $n \times n$ hermitian matrix. Show that

$$U := \exp(iK)$$

is a unitary matrix.

Solution 18.   Using $K = K^*$ we obtain from $U = \exp(iK)$ that $U^* = \exp(-iK)$. Then

$$U^*U = e^{-iK}e^{iK} = I_n .$$

Problem 19.   Let

$$A = \begin{pmatrix} 2 & 3 \\ 7 & -2 \end{pmatrix} .$$

Calculate $\det e^A$.

Solution 19.   To calculate $\det e^A$ would be quite clumsy. We use the identity

$$\det \exp(A) \equiv \exp(\operatorname{tr} A) .$$

Since $\operatorname{tr} A = 0$ we have

$$\det \exp(A) = e^0 = 1 .$$

Problem 20.   Let A be an $n \times n$ matrix over $\mathbf{C}$. Assume that $A^2 = cI_n$, where $c \in \mathbf{R}$. Calculate $\exp(A)$.

Solution 20.   Straightforward calculation yields

$$\exp(A) = I_n \cosh(\sqrt{c}) + \frac{A}{\sqrt{c}} \sinh(\sqrt{c}) .$$

Problem 21.   Let A be an $n \times n$ matrix with $A^3 = -A$ and $\mu \in \mathbf{R}$. Calculate $\exp(\mu A)$.

Solution 21. From $A^3 = -A$ it follows that

$$A^{2k+1} = (-1)^k A, \qquad k = 1, 2, \dots .$$

Thus

$$\exp(\mu A) = I_n + \sum_{k=1}^{\infty} \frac{\mu^k A^k}{k!}$$

$$= I_n + \sum_{k=0}^{\infty} \frac{\mu^{2k+1} A^{2k+1}}{(2k+1)!} + \sum_{k=1}^{\infty} \frac{\mu^{2k} A^{2k}}{(2k)!}$$

$$= I_n + \sum_{k=0}^{\infty} \frac{\mu^{2k+1}(-1)^k A}{(2k+1)!} - \sum_{k=1}^{\infty} \frac{\mu^{2k}(-1)^k A^2}{(2k)!}$$

$$= I_n + A \sin(\mu) + A^2(1 - \cos(\mu)) .$$

Problem 22. Let X be an $n \times n$ matrix over $\mathbf{C}$. Assume that $X^2 = I_n$. Let Y be an arbitrary $n \times n$ matrix over $\mathbf{C}$. Let $z \in \mathbf{C}$.
(i) Calculate $\exp(zX)Y\exp(-zX)$ using the *Baker-Campbell-Hausdorff relation*

$$e^{zX}Ye^{-zX} = Y + z[X, Y] + \frac{z^2}{2!}[X, [X, Y]] + \frac{z^3}{3!}[X, [X, [X, Y]]] + \cdots .$$

(ii) Calculate $\exp(zX)Y\exp(-zX)$ by first calculating $\exp(zX)$ and $\exp(-zX)$ and then doing the matrix multiplication. Compare the two methods.

Solution 22. (i) Using $X^2 = I_n$ we find for the first three commutators

$$[X, [X, Y]] = [X, XY - YX] = 2(Y - XYX)$$
$$[X, [X, [X, Y]]] = 2^2[X, Y]$$
$$[X, [X, [X, [X, Y]]]] = 2^3(Y - XYX) .$$

If the number of X's in the commutator is even (say, m, $m \geq 2$) we have

$$[X, [X, \dots [X, Y] \dots]] = 2^{m-1}(Y - XYX) .$$

If the number of X's in the commutator is odd (say m, $m \geq 3$) we have

$$[X, [X, \dots [X, Y] \dots]] = 2^{m-1}[X, Y] .$$

Thus

$$e^{zX}Ye^{-zX} = Y\left(1 + \frac{2^1 z^2}{2!} + \frac{2^3 z^4}{4!} + \cdots\right) + [X, Y]\left(z + \frac{2^2 z^3}{3!} + \cdots\right)$$

$$- XYX\left(\frac{2^1 z^2}{2!} + \frac{2^3 z^4}{4!} + \cdots\right) .$$

Consequently

$$e^{zX}Ye^{-zX} = Y\cosh^2(z) + [X,Y]\sinh(z)\cosh(z) - XYX\sinh^2(z).$$

(ii) Using the expansion

$$e^{zX} = \sum_{j=0}^{\infty} \frac{(zX)^j}{j!}$$

and $X^2 = I_n$ we have

$$e^{zX} = I_n\cosh(z) + X\sinh(z)$$
$$e^{-zX} = I_n\cosh(z) - X\sinh(z).$$

Matrix multplication yields

$$e^{zX}Ye^{-zX} = Y\cosh^2(z) + [X,Y]\sinh(z)\cosh(z) - XYX\sinh^2(z).$$

Problem 23. We consider the *principal logarithm* of a matrix $A \in \mathbf{C}^{n\times n}$ with no eigenvalues on $\mathbf{R}^-$ (the closed negative real axis). This logarithm is denoted by $\log A$ and is the unique matrix B such that $\exp(B) = A$ and the eigenvalues of B have imaginary parts lying strictly between $-\pi$ and π. For $A \in \mathbf{C}^{n\times n}$ with no eigenvalues on $\mathbf{R}^-$ we have the following integral representation

$$\log(s(A - I_n) + I_n) = \int_0^s (A - I_n)(t(A - I_n) + I_n)^{-1}dt.$$

Thus with $s = 1$ we obtain

$$\log A = \int_0^1 (A - I_n)(t(A - I_n) + I_n)^{-1}dt$$

where I_n is the $n \times n$ identity matrix. Let $A = xI_n$ with x a positive real number. Calculate $\log A$.

Solution 23. We have

$$\log(xI_n) = \int_0^1 (xI_n - I)(t(xI_n - I_n) + I_n)^{-1}dt$$
$$= \int_0^1 \frac{I_n(x - 1)}{I_n(t(x - 1) + 1)}dt$$
$$= I_n \int_0^1 \frac{(x - 1)}{t(x - 1) + 1}dt.$$

Setting $y = x - 1$ we obtain

$$\log(xI_n) = I_n \int_0^1 \frac{y}{ty+1} dt$$
$$= I_n y \int_0^1 \frac{1}{1+yt} dt$$
$$= I_n y \left. \frac{\log(1+yt)}{y} \right|_{t=0}^{t=1}$$
$$= I_n \log(1+y)$$
$$= I_n \log(x) .$$

Problem 24. Let A be a real or complex $n \times n$ matrix with no eigenvalues on $\mathbf{R}^-$ (the closed negative real axis). Then there exists a unique matrix X such that
1) $e^X = A$
2) the eigenvalues of X lie in the strip $\{ z : -\pi < \Im(z) < \pi \}$. We refer to X as the *principal logarithm* of A and write $X = \log A$. Similarly, there is a unique matrix S such that
1) $S^2 = A$
2) the eigenvalues of S lie in the open halfplane: $0 < \Re(z)$. We refer to S as the *principal square root* of A and write $S = A^{1/2}$.
If the matrix A is real then its principal logarithm and principal square root are also real.
The open halfplane associated with $z = \rho e^{i\theta}$ is the set of complex numbers $w = \zeta e^{i\phi}$ such that $-\pi/2 < \phi - \theta < \pi/2$.
Suppose that $A = BC$ has no eigenvalues on $\mathbf{R}^-$ and
1. $BC = CB$
2. every eigenvalue of B lies in the open halfplane of the corresponding eigenvalue of $A^{1/2}$ (or, equivalently, the same condition holds for C).

Show that
$$\log A = \log B + \log C .$$

Solution 24. First we show that the logarithms of B and C are well defined. Since $A = BC = CB$ it follows that A commutes with B and C. Thus we have $c = ab$ for the eigenvalue a of A, eigenvalue b of B and eigenvalue c of C. We express these eigenvalues in polar form as

$$a = \alpha e^{i\theta}, \quad b = \beta e^{i\phi}, \quad c = \gamma e^{i\psi} .$$

Since A has no eigenvalues on $\mathbf{R}^-$ we find that $-\pi < \theta < \pi$. The eigenvalues of B lie in the open halfplanes of the corresponding eigenvalues of $A^{1/2}$, that is

$$-\frac{\pi}{2} < \phi - \frac{\theta}{2} < \frac{\pi}{2}. \tag{1}$$

The relation $a = bc$ gives $\theta = \phi + \psi$, from which we have $\psi - \theta/2 = \theta/2 - \phi$. It follows from (1) that the eigenvalues of C lie in the open halfplanes of the corresponding eigenvalues of $A^{1/2}$. Thus, using $-\pi < \theta < \pi$, we obtain that B and C have no eigenvalues on $\mathbf{R}^-$ and their logarithms are well-defined. Next, we show that

$$e^{\log B + \log C} = A.$$

The matrices $\log B$ and $\log C$ commute since B and C do. Using the result that the exponential of the sum of commuting matrices is the product of the exponentials, we find

$$e^{\log B + \log C} = e^{\log B} e^{\log C} = BC = A.$$

It remains to show that the eigenvalues of $\log B + \log C$ have imaginary parts in $(-\pi, \pi)$. This follows since $BC = CB$ and therefore the eigenvalues of $\log B + \log C$ are $\log b + \log c = \log a$. Note that for $A = BC$ the commutativity condition $BC = CB$ is not enough to guarantee that $\log A = \log B + \log C$, as the following scalar example shows. Let $a = e^{-2\epsilon i}$ and $b = c = e^{(\pi - \epsilon)i}$ for small ϵ and positive. Then $a = bc$ but

$$\log a = -2\epsilon i \neq (\pi - \epsilon)i + (\pi - \epsilon)i = \log b + \log c.$$

This is due to the fact that b and c are equal to a nonprincipal square root of a, and hence are not in the halfplane of $a^{1/2}$.

Problem 25.   Let K be a hermitian matrix. Then $U := \exp(iK)$ is a unitary matrix. A method to find the hermitian matrix K from the unitary matrix U is to consider the principal logarithm of a matrix $A \in \mathbf{C}^{n \times n}$ with no eigenvalues on $\mathbf{R}^-$ (the closed negative real axis). This logarithm is denoted by $\log A$ and is the unique matrix B such that $\exp(B) = A$ and the eigenvalues of B have imaginary parts lying strictly between $-\pi$ and π. For $A \in \mathbf{C}^{n \times n}$ with no eigenvalues on $\mathbf{R}^-$ we have the following integral representation

$$\log(s(A - I_n) + I_n) = \int_0^s (A - I_n)(t(A - I_n) + I_n)^{-1} dt.$$

Thus with $s = 1$ we obtain

$$\log A = \int_0^1 (A - I_n)(t(A - I_n) + I_n)^{-1} dt$$

where I_n is the $n \times n$ identity matrix. Find $\log U$ of the unitary matrix

$$U = \frac{1}{\sqrt{2}} \begin{pmatrix} 1 & -1 \\ 1 & 1 \end{pmatrix}.$$

First test whether the method can be applied.

Solution 25.   We calculate $\log U$ to find iK given by $U = \exp(iK)$. We set $B = iK$ in the following. The eigenvalues of U are given by

$$\lambda_1 = \frac{1}{\sqrt{2}}(1+i), \qquad \lambda_2 = \frac{1}{\sqrt{2}}(1-i).$$

Thus the condition to apply the integral representation is satisfied. We consider first the general case $U = (u_{jk})$ and then simplify to $u_{11} = u_{22} = 1/\sqrt{2}$ and $u_{21} = -u_{12} = 1/\sqrt{2}$. We obtain

$$t(U - I_2) + I_2 = \begin{pmatrix} 1 + t(u_{11} - 1) & tu_{12} \\ tu_{21} & 1 + t(u_{22} - 1) \end{pmatrix}$$

and

$$d(t) := \det(t(U - I_2) + I_2) = 1 + t(-2 + \operatorname{tr}U) + t^2(1 - \operatorname{tr}U + \det U).$$

Let $X \equiv \det U - \operatorname{tr}U + 1$. Then

$$(U - I_2)(t(U - I_2) + I_2)^{-1} = \frac{1}{d(t)}\begin{pmatrix} tX + u_{11} - 1 & u_{12} \\ u_{21} & tX + u_{22} - 1 \end{pmatrix}.$$

With $u_{11} = u_{22} = 1/\sqrt{2}$, $u_{21} = -u_{12} = 1/\sqrt{2}$ we obtain

$$d(t) = 1 + t(-2 + \sqrt{2}) + t^2(2 - \sqrt{2})$$

and $X = 2 - \sqrt{2}$. Thus the matrix takes the form

$$\frac{1}{d(t)}\begin{pmatrix} t(2 - \sqrt{2}) + 1/\sqrt{2} - 1 & -1/\sqrt{2} \\ 1/\sqrt{2} & t(2 - \sqrt{2}) + 1/\sqrt{2} - 1 \end{pmatrix}.$$

Since

$$\int_0^1 \frac{1}{d(t)}dt = \frac{2}{\sqrt{2}}\left|\arctan\left(\frac{2(2 - \sqrt{2})t + \sqrt{2} - 2}{\sqrt{2}}\right)\right|_0^1 = \sqrt{2}\frac{\pi}{4}$$

and

$$\int_0^1 \frac{t}{d(t)}dt = \frac{1}{\sqrt{2}}\frac{\pi}{4}$$

we obtain

$$K = \begin{pmatrix} 0 & i\pi/4 \\ -i\pi/4 & 0 \end{pmatrix}.$$

Problem 26.   Let $\mathbf{x}_0, \mathbf{x}_1, \dots, \mathbf{x}_{2^n-1}$ be an orthonormal basis in $\mathbf{C}^{2^n}$. We define

$$U := \frac{1}{\sqrt{2^n}}\sum_{j=0}^{2^n-1}\sum_{k=0}^{2^n-1} e^{-i2\pi kj/2^n}\mathbf{x}_k\mathbf{x}_j^*. \tag{1}$$

Show that U is unitary. In other words show that $UU^* = I_{2^n}$, using the *completeness relation*

$$I_{2^n} = \sum_{j=0}^{2^n-1} \mathbf{x}_j \mathbf{x}_j^* .$$

Thus I_{2^n} is the $2^n \times 2^n$ unit matrix.

Solution 26.   From the definition (1) we find

$$U^* = \frac{1}{\sqrt{2^n}} \sum_{j=0}^{2^n-1} \sum_{k=0}^{2^n-1} e^{i2\pi kj/2^n} \mathbf{x}_j \mathbf{x}_k^*$$

where $*$ denotes the adjoint. Therefore

$$UU^* = \frac{1}{2^n} \sum_{j=0}^{2^n-1} \sum_{k=0}^{2^n-1} \sum_{l=0}^{2^n-1} \sum_{m=0}^{2^n-1} e^{i2\pi(kj-lm)/2^n} \mathbf{x}_j \mathbf{x}_k^* \mathbf{x}_l \mathbf{x}_m^*$$

$$= \frac{1}{2^n} \sum_{j=0}^{2^n-1} \sum_{k=0}^{2^n-1} \sum_{m=0}^{2^n-1} e^{i2\pi(kj-km)/2^n} \mathbf{x}_j \mathbf{x}_m^* .$$

We have for $j = m$, $e^{i2\pi(kj-km)/2^n} = 1$. Thus for $j, m = 0, 1, \ldots, 2^n - 1$

$$\sum_{k=0}^{2^n-1} (e^{i2\pi(j-m)/2^n})^k = 2^n, \quad j = m$$

$$\sum_{k=0}^{2^n-1} (e^{i2\pi(j-m)/2^n})^k = \frac{1 - e^{i2\pi(j-m)}}{1 - e^{i2\pi(j-m)/2^n}} = 0, \quad j \neq m.$$

Thus

$$UU^* = \sum_{j=0}^{2^n-1} \mathbf{x}_j \mathbf{x}_j = I_{2^n} .$$

Problem 27.   Consider the unitary matrix

$$U = \begin{pmatrix} 0 & 1 \\ 1 & 0 \end{pmatrix} .$$

Show that we can find a unitary matrix V such that $V^2 = U$. Thus V would be the square root of U. What are the eigenvalues of V?

Solution 27.   We find the unitary matrix

$$V = \frac{1}{2} \begin{pmatrix} 1+i & 1-i \\ 1-i & 1+i \end{pmatrix} .$$

Obviously $-V$ is also a square root. The eigenvalues of V are 1 and i. The eigenvalues of $-V$ are -1 and $-i$. Note that the eigenvalues of U are 1 and -1.

Problem 28. Let A be an $n \times n$ matrix. Let $\omega, \mu \in \mathbf{R}$. Assume that

$$\|e^{tA}\| \le Me^{\omega t}, \qquad t \ge 0$$

and $\mu > \omega$. Then we have

$$(\mu I_n - A)^{-1} \equiv \int_0^\infty e^{-\mu t} e^{tA} dt . \tag{1}$$

Calculate the left and right-hand side of (1) for the matrix

$$A = \begin{pmatrix} 0 & 1 \\ 1 & 0 \end{pmatrix} .$$

Solution 28. For the left-hand side we have

$$(\mu I_2 - A)^{-1} = \begin{pmatrix} \mu & -1 \\ -1 & \mu \end{pmatrix}^{-1} = \frac{1}{1 - \mu^2} \begin{pmatrix} -\mu & 1 \\ -1 & -\mu \end{pmatrix} .$$

Since

$$\exp(tA) = I_2 \cosh(t) + \begin{pmatrix} 0 & 1 \\ 1 & 0 \end{pmatrix} \sinh(t) = \begin{pmatrix} \cosh(t) & \sinh(t) \\ \sinh(t) & \cosh(t) \end{pmatrix}$$

and

$$\exp(-\mu t I_2) = \begin{pmatrix} e^{-\mu t} & 0 \\ 0 & e^{-\mu t} \end{pmatrix}$$

we find for the right-hand side

$$\int_0^\infty e^{-\mu t} e^{tA} dt = \int_0^\infty e^{-\mu t I_2} e^{tA} dt$$

$$= \int_0^\infty \begin{pmatrix} e^{-\mu t} & 0 \\ 0 & e^{-\mu t} \end{pmatrix} \begin{pmatrix} \cosh(t) & \sinh(t) \\ \sinh(t) & \cosh(t) \end{pmatrix} dt$$

$$= \int_0^\infty \begin{pmatrix} e^{-\mu t} \cosh(t) & e^{-\mu t} \sinh(t) \\ e^{-\mu t} \sinh(t) & e^{-\mu t} \cosh(t) \end{pmatrix} dt .$$

Since

$$\int_0^\infty \cosh(t) e^{-\mu t} dt = -\frac{\mu}{1 - \mu^2}$$

and

$$\int_0^\infty \sinh(t) e^{-\mu t} dt = -\frac{1}{1 - \mu^2}$$

we obtain

$$\int_0^\infty e^{-\mu t} e^{tA} dt = \frac{1}{1-\mu^2} \begin{pmatrix} -\mu & -1 \\ -1 & -\mu \end{pmatrix}.$$

Problem 29. The *Fréchet derivative* of a matrix function $f : \mathbf{C}^{n\times n}$ at a point $X \in \mathbf{C}^{n\times n}$ is a linear mapping $L_X : \mathbf{C}^{n\times n} \to \mathbf{C}^{n\times n}$ such that for all $Y \in \mathbf{C}^{n\times n}$

$$f(X+Y) - f(X) - L_X(Y) = o(\|Y\|).$$

Calculate the Fréchet derivative of $f(X) = X^2$.

Solution 29. We have

$$f(X+Y) - f(X) = XY + YX + Y^2.$$

Thus

$$L_X(Y) = XY + YX.$$

The right-hand side is the anti-commutator of X and Y.

Chapter 8

Linear Differential Equations

Let A be an $n \times n$ matrix over $\mathbf{R}$. Homogeneous systems of linear differential equations with constant coefficients are given by

$$\frac{d\mathbf{x}}{dt} = A\mathbf{x}, \qquad \mathbf{x} = (x_1, x_2, \ldots, x_n)^T \tag{1}$$

with the initial condition $\mathbf{x}(t = 0) = \mathbf{x}_0$. The solution is given by

$$\mathbf{x}(t) = \exp(tA)\mathbf{x}_0 . \tag{2}$$

The nonhomogeneous system of linear differential equations with constant coefficients are given by

$$\frac{d\mathbf{x}}{dt} = A\mathbf{x} + \mathbf{y}(t) \tag{3}$$

where $y_1(t), y_2(t), \ldots, y_n(t)$ are analytic functions. The solution can be found with the method called *variation of constants*. The solution which vanishes at $t = 0$ is given by

$$\mathbf{x}(t) = \int_0^t \exp((t - \tau)A)\mathbf{y}(\tau)d\tau = \exp(tA) \int_0^t \exp(-\tau A)\mathbf{y}(\tau)d\tau . \tag{4}$$

Thus the solution of a system of differential equations (3) is the sum of the general solution of (1) given by (2) and the particular solution given by (4).

Problem 1.   Solve the initial value problem of the linear differential equation

$$\frac{dx}{dt} = 2x + \sin t.$$

Solution 1.   The general solution of the homogeneous part is

$$x(t) = e^{2t}x_0.$$

The particular solution of the inhomogeneous part is

$$\int_0^t \exp(2(t-\tau))\sin\tau\, d\tau = e^{2t}\int_0^t \exp(-2\tau)\sin\tau\, d\tau$$

$$= -\frac{2\sin t + \cos t}{5} + \frac{e^{2t}}{5}.$$

Problem 2.   Solve the initial value problem of $dx/dt = Ax$, where

$$A = \begin{pmatrix} 0 & 1 \\ 1 & 0 \end{pmatrix}.$$

Solution 2.   Since

$$\exp(tA) = \begin{pmatrix} \cosh(t) & \sinh(t) \\ \sinh(t) & \cosh(t) \end{pmatrix}$$

we obtain

$$\begin{pmatrix} x_1(t) \\ x_2(t) \end{pmatrix} = \exp(tA)\begin{pmatrix} x_1(0) \\ x_2(0) \end{pmatrix} = \begin{pmatrix} x_1(0)\cosh(t) + x_2(0)\sinh(t) \\ x_1(0)\sinh(t) + x_2(0)\cosh(t) \end{pmatrix}.$$

Problem 3.   Solve the initial value problem of $dx/dt = Ax$, where

$$A = \begin{pmatrix} a & c \\ 0 & b \end{pmatrix}, \quad a, b, c \in \mathbf{R}.$$

Solution 3.   We have

$$\exp(tA) = \begin{pmatrix} e^{at} & dt \\ 0 & e^{bt} \end{pmatrix}$$

where

$$d = \frac{c(e^{bt} - e^{at})}{b - a} \quad \text{for} \quad b \neq a$$

and

$$d = ce^{at} \quad \text{for} \quad b = a .$$

Thus

$$\begin{pmatrix} x_1(t) \\ x_2(t) \end{pmatrix} = \begin{pmatrix} e^{at} & dt \\ 0 & e^{bt} \end{pmatrix} \begin{pmatrix} x_1(0) \\ x_2(0) \end{pmatrix} .$$

Problem 4. Show that the n-th order differential equation

$$\frac{d^n x}{dt^n} = c_0 x + c_1 \frac{dx}{dt} + \cdots + c_{n-1} \frac{d^{n-1} x}{dt^{n-1}}, \quad c_j \in \mathbf{R}$$

can be written as a system of first order differential equation.

Solution 4. We set $x_1 = x$ and

$$\frac{dx_1}{dt} := x_2, \quad \frac{dx_2}{dt} := x_3, \quad \ldots \quad , \frac{dx_n}{dt} := c_0 x_1 + c_1 x_2 + \cdots + c_{n-1} x_n .$$

Thus we have the system $dx/dt = Ax$ with the matrix

$$A = \begin{pmatrix} 0 & 1 & 0 & \cdots & 0 \\ 0 & 0 & 1 & \cdots & 0 \\ . & . & . & \cdots & . \\ 0 & 0 & 0 & \cdots & 1 \\ c_0 & c_1 & c_2 & \cdots & c_{n-1} \end{pmatrix} .$$

Problem 5. Let A, X, F be $n \times n$ matrices. Assume that the matrix elements of X and F are differentiable functions of t. Consider the initial-value linear matrix differential equation with an inhomogeneous part

$$\frac{dX(t)}{dt} = AX(t) + F(t), \quad X(t_0) = C .$$

Find the solution of this matrix differential equation.

Solution 5. The solution is given by

$$X(t) = e^{(t-t_0)A} C + e^{tA} \int_{t_0}^{t} e^{-sA} F(s) ds$$

or, equivalently

$$X(t) = e^{(t-t_0)A} C + \int_{t_0}^{t} e^{(t-s)A} F(s) ds .$$

Problem 6. Let A, B, C, Y be $n \times n$ matrices. We know that

$$AY + YB = C$$

can be written as

$$((I_n \otimes A) + (B^T \otimes I_n))\text{vec}Y = \text{vec}C$$

where $\otimes$ denotes the Kronecker product. The *vec operation* is defined as

$$\text{vec}Y := (y_{11}, \ldots, y_{n1}, y_{12}, \ldots, y_{n2}, \ldots, y_{1n}, \ldots, y_{nn})^T.$$

Apply the vec operation to the matrix differential equation

$$\frac{d}{dt}X(t) = AX(t) + X(t)B$$

where A, B are $n \times n$ matrices and the initial matrix $X(t = 0) \equiv X(0)$ is given. Find the solution of this differential equation.

Solution 6. The vec operation is linear and using the result from above we obtain

$$\frac{d}{dt}\text{vec}X(t) = (I_n \otimes A + B^T \otimes I_n)\text{vec}X(t).$$

The solution of the linear differential equation is given by

$$\text{vec}X(t) = e^{t(I_n \otimes A + B^T \otimes I_n)}\text{vec}X(0).$$

Since $[I_n \otimes A, B^T \otimes I_n] = 0_{n^2}$ this can be written as

$$\text{vec}X(t) = (e^{tB^T} \otimes e^{tA})\text{vec}X(0).$$

Problem 7. The *motion of a charge q* in an electromagnetic field is given by

$$m\frac{d\mathbf{v}}{dt} = q(\mathbf{E} + \mathbf{v} \times \mathbf{B}) \tag{1}$$

where m denotes the mass and $\mathbf{v}$ the velocity. Assume that

$$\mathbf{E} = \begin{pmatrix} E_1 \\ E_2 \\ E_3 \end{pmatrix}, \qquad \mathbf{B} = \begin{pmatrix} B_1 \\ B_2 \\ B_3 \end{pmatrix} \tag{2}$$

are constant fields. Find the solution of the initial value problem.

Solution 7. Equation (1) can be written in the form

$$\begin{pmatrix} dv_1/dt \\ dv_2/dt \\ dv_3/dt \end{pmatrix} = \frac{q}{m}\begin{pmatrix} 0 & B_3 & -B_2 \\ -B_3 & 0 & B_1 \\ B_2 & -B_1 & 0 \end{pmatrix}\begin{pmatrix} v_1 \\ v_2 \\ v_3 \end{pmatrix} + \frac{q}{m}\begin{pmatrix} E_1 \\ E_2 \\ E_3 \end{pmatrix}. \tag{3}$$

We set

$$B_j \to \frac{q}{m} B_j, \qquad E_j \to \frac{q}{m} E_j. \tag{4}$$

Thus

$$\begin{pmatrix} dv_1/dt \\ dv_2/dt \\ dv_3/dt \end{pmatrix} = \begin{pmatrix} 0 & B_3 & -B_2 \\ -B_3 & 0 & B_1 \\ B_2 & -B_1 & 0 \end{pmatrix} \begin{pmatrix} v_1 \\ v_2 \\ v_3 \end{pmatrix} + \begin{pmatrix} E_1 \\ E_2 \\ E_3 \end{pmatrix}. \tag{5}$$

Equation (5) is a system of nonhomogeneous linear differential equations with constant coefficients. The solution to the homogeneous equation

$$\begin{pmatrix} dv_1/dt \\ dv_2/dt \\ dv_3/dt \end{pmatrix} = \begin{pmatrix} 0 & B_3 & -B_2 \\ -B_3 & 0 & B_1 \\ B_2 & -B_1 & 0 \end{pmatrix} \begin{pmatrix} v_1 \\ v_2 \\ v_3 \end{pmatrix} \tag{6}$$

is given by

$$\begin{pmatrix} v_1(t) \\ v_2(t) \\ v_3(t) \end{pmatrix} = e^{tM} \begin{pmatrix} v_1(0) \\ v_2(0) \\ v_3(0) \end{pmatrix} \tag{7}$$

where M is the matrix of the right-hand side of (6) and $v_j(0) = v_j(t = 0)$. The solution of the system of nonhomogeneous linear differential equations (5) can be found with the help of the method called *variation of constants*. One sets

$$\mathbf{v}(t) = e^{tM} \mathbf{f}(t) \tag{8}$$

where $\mathbf{f} : \mathbf{R} \to \mathbf{R}^3$ is some differentiable curve. Then

$$\frac{d\mathbf{v}}{dt} = M e^{tM} \mathbf{f}(t) + e^{tM} \frac{d\mathbf{f}}{dt}. \tag{9}$$

Inserting (9) into (5) yields

$$M\mathbf{v}(t) + \mathbf{E} = M e^{tM} \mathbf{f}(t) + e^{tM} \frac{d\mathbf{f}}{dt} = M\mathbf{v}(t) + e^{tM} \frac{d\mathbf{f}}{dt}. \tag{10}$$

Consequently

$$\frac{d\mathbf{f}}{dt} = e^{-tM} \mathbf{E}. \tag{11}$$

By integration we obtain

$$\mathbf{f}(t) = \int_0^t e^{-sM} \mathbf{E} ds + \mathbf{K} \tag{12}$$

where

$$\mathbf{K} = \begin{pmatrix} K_1 \\ K_2 \\ K_3 \end{pmatrix}. \tag{13}$$

Therefore we obtain the general solution of the initial value problem of the nonhomogeneous system (5), namely

$$\mathbf{v}(t) = e^{tM} \left(\int_0^t e^{-sM} \mathbf{E} ds + \mathbf{K} \right) \tag{14}$$

where

$$\mathbf{K} = \begin{pmatrix} v_1(0) \\ v_2(0) \\ v_3(0) \end{pmatrix}. \tag{15}$$

We now have to calculate e^{tM} and e^{-sM}. We find that

$$M^2 = \begin{pmatrix} -B_2^2 - B_3^2 & B_1 B_2 & B_1 B_3 \\ B_1 B_2 & -B_1^2 - B_3^2 & B_2 B_3 \\ B_1 B_3 & B_2 B_3 & -B_1^2 - B_2^2 \end{pmatrix} \tag{16}$$

and

$$M^3 = M^2 M = -B^2 M \tag{17}$$

where

$$B^2 = B_1^2 + B_2^2 + B_3^2 \tag{18}$$

and

$$B := \sqrt{B_1^2 + B_2^2 + B_3^2}.$$

Therefore $M^4 = -B^2 M^2$. Since

$$e^{tM} := \sum_{k=0}^{\infty} \frac{(tM)^k}{k!} = I + \frac{tM}{1!} + \frac{t^2 M^2}{2!} + \frac{t^3 M^3}{3!} + \frac{t^4 M^4}{4!} + \cdots \tag{19}$$

where I denotes the 3×3 unit matrix, we obtain, taking (17) into account

$$e^{tM} = I + M \left(t - \frac{t^3}{3!} B^2 + \frac{t^5}{5!} B^4 - \cdots \right) + M^2 \left(\frac{t^2}{2!} - \frac{t^4}{4!} B^2 + \frac{t^6}{6!} B^4 - \cdots \right).$$

Thus

$$e^{tM} = I + \frac{M}{B} \left(tB - \frac{t^3}{3!} BB^2 + \cdots \right) - \frac{M^2}{B^2} \left(1 - 1 - \frac{t^2 B^2}{2!} + \frac{t^4 B^4}{4!} - \cdots \right)$$

$$= I + \frac{M}{B} \sin(Bt) + \frac{M^2}{B^2} (1 - \cos(Bt)). \tag{20}$$

Therefore

$$e^{tM} = I + \frac{M}{B} \sin(Bt) + \frac{M^2}{B^2} (1 - \cos(Bt)) \tag{21}$$

and

$$e^{-sM} = I - \frac{M}{B} \sin(Bs) + \frac{M^2}{B^2} (1 - \cos(Bs)). \tag{22}$$

Equation (22) follows from (21) by replacing $t \to -s$. Since

$$\int_0^t e^{-sM} \mathbf{E} ds = \int_0^t \left(I - \frac{M}{B} \sin(Bs) + \frac{M^2}{B^2}(1 - \cos(Bs)) \right) \mathbf{E} ds$$

$$= \mathbf{E}t + \frac{M\mathbf{E}}{B^2} \cos(Bt) - \frac{M\mathbf{E}}{B^2} + \frac{M^2 \mathbf{E}t}{B^2} - \frac{M^2 \mathbf{E}}{B^2 B} \sin(Bt)$$

we find as the solution of the initial value problem of system (5)

$$\mathbf{v}(t) = \mathbf{E}t \left(1 + \frac{M^2}{B^2} \right) - \frac{M^2 \mathbf{E}}{B^2 B} \sin(Bt) + \frac{M\mathbf{v}(0)}{B} \sin(Bt)$$

$$+ \frac{M\mathbf{E}}{B^2}(1 - \cos(Bt)) + \frac{M^2 \mathbf{v}(0)}{B^2}(1 - \cos(Bt)) + \mathbf{v}(0)$$

where $\mathbf{v}(t = 0) \equiv \mathbf{v}(0)$.

Problem 8. Consider a system of linear ordinary differential equations with periodic coefficients

$$\frac{d\mathbf{x}}{dt} = A(t)\mathbf{x} \tag{1}$$

where $A(t)$ is an $n \times n$ matrix of periodic functions with a period T. From *Floquet theory* we know that any fundamental $n \times n$ matrix $\Phi(t)$, which is defined as a nonsingular matrix satisfying the matrix differential equation

$$\frac{d\Phi(t)}{dt} = A(t)\Phi(t)$$

can be expressed as

$$\Phi(t) = P(t) \exp(tR). \tag{2}$$

Here $P(t)$ is a nonsingular $n \times n$ matrix of periodic functions with the same period T, and R, a constant matrix, whose eigenvalues are called the *characteristic exponents* of the periodic system (1). Let

$$\mathbf{y} = P^{-1}(t)\mathbf{x}.$$

Show that $\mathbf{y}$ satisfies the system of linear differential equations with constant coefficients

$$\frac{d\mathbf{y}}{dt} = R\mathbf{y}.$$

Solution 8. From $P(t)P^{-1}(t) = I_n$ we have

$$\frac{dP^{-1}(t)}{dt} = -P^{-1}(t)\frac{dP(t)}{dt}P^{-1}(t).$$

and from (2) it follows that

$$\frac{d\Phi(t)}{dt} = \frac{dP(t)}{dt}e^{tR} + P(t)Re^{tR}$$

or

$$\frac{dP(t)}{dt} = \frac{d\Phi(t)}{dt}e^{-tR} - P(t)R.$$

Using these two results we obtain from $\mathbf{y} = P^{-1}(t)\mathbf{x}$

$$\frac{d\mathbf{y}}{dt} = \frac{dP^{-1}(t)}{dt}\mathbf{x} + P^{-1}(t)\frac{d\mathbf{x}}{dt}$$

$$= -P^{-1}(t)\frac{dP(t)}{dt}P^{-1}(t)\mathbf{x} + P^{-1}(t)A(t)\mathbf{x}$$

$$= -P^{-1}(t)\left(\frac{d\Phi(t)}{dt}e^{-tR} - P(t)R\right)\mathbf{y} + P^{-1}(t)A(t)\mathbf{x}$$

$$= -P^{-1}(t)\frac{d\Phi(t)}{dt}e^{-tR}\mathbf{y} + R\mathbf{y} + P^{-1}(t)A(t)P(t)\mathbf{y}$$

$$= -P^{-1}(t)A(t)\Phi(t)e^{-tR}\mathbf{y} + P^{-1}(t)A(t)P(t) + R\mathbf{y}$$

$$= P^{-1}(t)A(t)(-\Phi(t)e^{-tR} + P(t))\mathbf{y} + R\mathbf{y}$$

$$= R\mathbf{y}.$$

Problem 9.   Consider the autonomous system of nonlinear first order ordinary differential equations

$$\frac{dx_1}{dt} = a(x_2 - x_1) = f_1(x_1, x_2, x_3)$$

$$\frac{dx_2}{dt} = (c - a)x_1 + cx_2 - x_1x_3 = f_2(x_1, x_2, x_3)$$

$$\frac{dx_3}{dt} = -bx_3 + x_1x_2 = f_3(x_1, x_2, x_3)$$

where $a > 0$, $b > 0$ and c are real constants with $2c > a$.
(i) The *fixed points* are defined as the solutions of the system of equations

$$f_1(x_1^*, x_2^*, x_3^*) = a(x_2^* - x_1^*) = 0$$
$$f_2(x_1^*, x_2^*, x_3^*) = (c - a)x_1^* + cx_2^* - x_1^*x_3^* = 0$$
$$f_3(x_1^*, x_2^*, x_3^*) = -bx_3^* + x_1^*x_2^* = 0.$$

Find the fixed points. Obviously $(0, 0, 0)$ is a fixed point.
(ii) The *linearized equation* (or variational equation) is given by

$$\begin{pmatrix} dy_1/dt \\ dy_2/dt \\ dy_3/dt \end{pmatrix} = A \begin{pmatrix} y_1 \\ y_2 \\ y_3 \end{pmatrix}$$

where the 3×3 matrix A is given by

$$A_{\mathbf{x}=\mathbf{x}^*} = \begin{pmatrix} \partial f_1/\partial x_1 & \partial f_1/\partial x_2 & \partial f_1/\partial x_3 \\ \partial f_2/\partial x_1 & \partial f_2/\partial x_2 & \partial f_2/\partial x_3 \\ \partial f_3/\partial x_1 & \partial f_3/\partial x_2 & \partial f_3/\partial x_3 \end{pmatrix}_{\mathbf{x}=\mathbf{x}^*}$$

where $\mathbf{x} = \mathbf{x}^*$ indicates to insert one of the fixed points into A. Calculate A and insert the first fixed point $(0,0,0)$. Calculate the eigenvalues of A. If all eigenvalues have negative real part then the fixed point is stable. Thus study the stability of the fixed point.

Solution 9. (i) Obviously

$$x_1^* = x_2^* = x_3^* = 0$$

is a solution. The other two solutions we find by eliminating x_3^* from the third equation and x_2^* from the first equation. We obtain

$$x_1^* = \sqrt{b(2c - a)}, \quad x_2^* = \sqrt{b(2c - a)}, \quad x_3^* = 2c - a$$

and

$$x_1^* = -\sqrt{b(2c - a)}, \quad x_2^* = -\sqrt{b(2c - a)}, \quad x_3^* = 2c - a.$$

(ii) We obtain

$$\begin{pmatrix} \partial f_1/\partial x_1 & \partial f_1/\partial x_2 & \partial f_1/\partial x_3 \\ \partial f_2/\partial x_1 & \partial f_2/\partial x_2 & \partial f_2/\partial x_3 \\ \partial f_3/\partial x_1 & \partial f_3/\partial x_2 & \partial f_3/\partial x_3 \end{pmatrix} = \begin{pmatrix} -a & a & 0 \\ c - a - x_3 & c & -x_1 \\ x_2 & x_1 & -b \end{pmatrix}.$$

Inserting the fixed point $(0,0,0)$ yields the matrix

$$A_{\mathbf{x}^* = (0,0,0)} = \begin{pmatrix} -a & a & 0 \\ c - a & c & 0 \\ 0 & 0 & -b \end{pmatrix}.$$

The eigenvalues are

$$\lambda_{1,2} = -\frac{a - c}{2} \pm \sqrt{a(2c - a)}, \qquad \lambda_3 = -b.$$

Depending on c the fixed point $(0,0,0)$ is unstable.

Chapter 9

Kronecker Product

Let A be an $m \times n$ matrix and B be an $r \times s$ matrix. The *Kronecker product* of A and B is defined as the $(m \cdot r) \times (n \cdot s)$ matrix

$$A \otimes B := \begin{pmatrix} a_{11}B & a_{12}B & \dots & a_{1n}B \\ a_{21}B & a_{22}B & \dots & a_{2n}B \\ \vdots & \vdots & \ddots & \vdots \\ a_{m1}B & a_{m2}B & \dots & a_{mn}B \end{pmatrix}.$$

Thus $A \otimes B$ is an $mr \times ns$ matrix.

We have the properties: The Kronecker product is associative

$$(A \otimes B) \otimes C = A \otimes (B \otimes C).$$

Let A and B be $m \times n$ matrices and C a $p \times q$ matrix. Then

$$(A + B) \otimes C = A \otimes C + B \otimes C.$$

Let $c \in \mathbf{C}$. Then

$$((cA) \otimes B) = c(A \otimes B) = (A \otimes (cB)).$$

Let A be an $m \times n$ matrix, B be a $p \times q$ matrix, C be an $n \times r$ matrix and D be a $q \times s$ matrix. Then

$$(A \otimes B)(C \otimes D) = (AC) \otimes (BD).$$

Problem 1. (i) Let

$$\mathbf{x} = \frac{1}{\sqrt{2}} \begin{pmatrix} 1 \\ 1 \end{pmatrix}, \qquad \mathbf{y} = \frac{1}{\sqrt{2}} \begin{pmatrix} 1 \\ -1 \end{pmatrix}.$$

Thus $\{\mathbf{x}, \mathbf{y}\}$ forms an orthonormal basis in $\mathbf{C}^2$ (Hadamard basis). Calculate

$$\mathbf{x} \otimes \mathbf{x}, \quad \mathbf{x} \otimes \mathbf{y}, \quad \mathbf{y} \otimes \mathbf{x}, \quad \mathbf{y} \otimes \mathbf{y}$$

and interpret the result.

Solution 1. We obtain

$$\mathbf{x} \otimes \mathbf{x} = \frac{1}{2} \begin{pmatrix} 1 \\ 1 \\ 1 \\ 1 \end{pmatrix}, \qquad \mathbf{x} \otimes \mathbf{y} = \frac{1}{2} \begin{pmatrix} 1 \\ -1 \\ 1 \\ -1 \end{pmatrix}$$

$$\mathbf{y} \otimes \mathbf{x} = \frac{1}{2} \begin{pmatrix} 1 \\ 1 \\ -1 \\ -1 \end{pmatrix}, \qquad \mathbf{y} \otimes \mathbf{y} = \frac{1}{2} \begin{pmatrix} 1 \\ -1 \\ -1 \\ 1 \end{pmatrix}.$$

Thus we find an orthonormal basis in $\mathbf{C}^4$ from an orthonormal basis in $\mathbf{C}^2$.

Problem 2. Consider the Pauli matrices

$$\sigma_x := \begin{pmatrix} 0 & 1 \\ 1 & 0 \end{pmatrix}, \qquad \sigma_z := \begin{pmatrix} 1 & 0 \\ 0 & -1 \end{pmatrix}.$$

Find $\sigma_x \otimes \sigma_z$ and $\sigma_z \otimes \sigma_x$. Is $\sigma_x \otimes \sigma_z = \sigma_z \otimes \sigma_x$?

Solution 2. We obtain

$$\sigma_x \otimes \sigma_z = \begin{pmatrix} 0 & 0 & 1 & 0 \\ 0 & 0 & 0 & -1 \\ 1 & 0 & 0 & 0 \\ 0 & -1 & 0 & 0 \end{pmatrix}$$

and

$$\sigma_z \otimes \sigma_x = \begin{pmatrix} 0 & 1 & 0 & 0 \\ 1 & 0 & 0 & 0 \\ 0 & 0 & 0 & -1 \\ 0 & 0 & -1 & 0 \end{pmatrix}.$$

We see that $\sigma_x \otimes \sigma_z \neq \sigma_z \otimes \sigma_x$.

Problem 3.   Every 4×4 unitary matrix U can be written as

$$U = (U_1 \otimes U_2) \exp(i(\alpha\sigma_x \otimes \sigma_x + \beta\sigma_y \otimes \sigma_y + \gamma\sigma_z \otimes \sigma_z))(U_3 \otimes U_4)$$

where $U_j \in U(2)$ $(j = 1, 2, 3, 4)$ and $\alpha, \beta, \gamma \in \mathbf{R}$. Calculate

$$\exp(i(\alpha\sigma_x \otimes \sigma_x + \beta\sigma_y \otimes \sigma_y + \gamma\sigma_z \otimes \sigma_z)).$$

Solution 3.   Since

$$[\sigma_x \otimes \sigma_x, \sigma_y \otimes \sigma_y] = 0$$
$$[\sigma_x \otimes \sigma_x, \sigma_z \otimes \sigma_z] = 0$$
$$[\sigma_y \otimes \sigma_y, \sigma_z \otimes \sigma_z] = 0$$

we can write

$$\exp(i(\alpha\sigma_x \otimes \sigma_x + \beta\sigma_y \otimes \sigma_y + \gamma\sigma_z \otimes \sigma_z)) = e^{i\alpha\sigma_x \otimes \sigma_x} e^{i\beta\sigma_y \otimes \sigma_y} e^{i\gamma\sigma_z \otimes \sigma_z}.$$

Since $(j = x, y, z)$

$$(\sigma_j \otimes \sigma_j)(\sigma_j \otimes \sigma_j) = I_2 \otimes I_2$$

and

$$e^{i\alpha\sigma_x \otimes \sigma_x} = I_4 \cos\alpha + i(\sigma_x \otimes \sigma_x)\sin\alpha$$
$$e^{i\beta\sigma_y \otimes \sigma_y} = I_4 \cos\beta + i(\sigma_y \otimes \sigma_y)\sin\beta$$
$$e^{i\gamma\sigma_z \otimes \sigma_z} = I_4 \cos\gamma + i(\sigma_z \otimes \sigma_z)\sin\gamma$$

we find

$$e^{i\gamma}\begin{pmatrix} \cos(\alpha - \beta) & 0 & 0 & i\sin(\alpha - \beta) \\ 0 & e^{-2i\gamma}\cos(\alpha + \beta) & -ie^{-2i\gamma}\sin(\alpha + \beta) & 0 \\ 0 & -ie^{-2i\gamma}\sin(\alpha + \beta) & e^{-2i\gamma}\cos(\alpha + \beta) & 0 \\ i\sin(\alpha - \beta) & 0 & 0 & \cos(\alpha - \beta) \end{pmatrix}.$$

Problem 4.   Find an orthonormal basis given by hermitian matrices in the Hilbert space $\mathcal{H}$ of 4×4 matrices over $\mathbf{C}$. The *scalar product* in the Hilbert space $\mathcal{H}$ is given by

$$\langle A, B \rangle := \mathrm{tr}(AB^*), \qquad A, B \in \mathcal{H}.$$

Hint.   Start with hermitian 2×2 matrices and then use the Kronecker product.

Solution 4.   For 2×2 matrices an orthonormal basis of hermitian matrices is given by

$$\mu_0 = \frac{1}{\sqrt{2}}I_2, \qquad \mu_1 = \frac{1}{\sqrt{2}}\sigma_x = \frac{1}{\sqrt{2}}\begin{pmatrix} 0 & 1 \\ 1 & 0 \end{pmatrix}$$

$$\mu_2 = \frac{1}{\sqrt{2}}\sigma_y = \frac{1}{\sqrt{2}}\begin{pmatrix} 0 & -i \\ i & 0 \end{pmatrix}, \qquad \mu_3 = \frac{1}{\sqrt{2}}\sigma_z = \frac{1}{\sqrt{2}}\begin{pmatrix} 1 & 0 \\ 0 & -1 \end{pmatrix}.$$

This means μ_1, μ_2, μ_3 are rescaled Pauli spin matrices and μ_0 is the rescaled 2×2 identity matrix. Now the Kronecker product of two hermitian matrices is again hermitian. Thus forming all possible 16 Kronecker products

$$\mu_j \otimes \mu_k, \qquad j,k = 0,1,2,3$$

we find an orthonormal basis of the Hilbert space of the 4×4 matrices, where we used

$$\mathrm{tr}((\mu_j \otimes \mu_k)(\mu_m \otimes \mu_n)) = \mathrm{tr}((\mu_j\mu_m) \otimes (\mu_k\mu_n)) = (\mathrm{tr}(\mu_j\mu_m))(\mathrm{tr}(\mu_k\mu_n))$$

with $j,k,m,n = 0,1,2,3$. We applied that $\mathrm{tr}(A \otimes B) = \mathrm{tr}A\,\mathrm{tr}B$, where A and B are $n \times n$ matrices.

Problem 5. Consider the 4×4 matrices

$$\alpha_1 = \begin{pmatrix} 0 & 0 & 0 & 1 \\ 0 & 0 & 1 & 0 \\ 0 & 1 & 0 & 0 \\ 1 & 0 & 0 & 0 \end{pmatrix} = \sigma_x \otimes \sigma_x$$

$$\alpha_2 = \begin{pmatrix} 0 & 0 & 0 & -i \\ 0 & 0 & i & 0 \\ 0 & -i & 0 & 0 \\ i & 0 & 0 & 0 \end{pmatrix} = \sigma_x \otimes \sigma_y$$

$$\alpha_3 = \begin{pmatrix} 0 & 0 & 1 & 0 \\ 0 & 0 & 0 & -1 \\ 1 & 0 & 0 & 0 \\ 0 & -1 & 0 & 0 \end{pmatrix} = \sigma_x \otimes \sigma_z\,.$$

Let $\mathbf{a} = (a_1,a_2,a_3)$, $\mathbf{b} = (b_1,b_2,b_3)$, $\mathbf{c} = (c_1,c_2,c_3)$, $\mathbf{d} = (d_1,d_2,d_3)$ be elements in $\mathbf{R}^3$ and

$$\mathbf{a} \cdot \boldsymbol{\alpha} := a_1\alpha_1 + a_2\alpha_2 + a_3\alpha_3\,.$$

Calculate

$$\mathrm{tr}((\mathbf{a} \cdot \boldsymbol{\alpha})(\mathbf{b} \cdot \boldsymbol{\alpha}))$$

and

$$\mathrm{tr}((\mathbf{a} \cdot \boldsymbol{\alpha})(\mathbf{b} \cdot \boldsymbol{\alpha})(\mathbf{c} \cdot \boldsymbol{\alpha})(\mathbf{d} \cdot \boldsymbol{\alpha}))\,.$$

Solution 5. We find

$$\mathrm{tr}((\mathbf{a} \cdot \boldsymbol{\alpha})(\mathbf{b} \cdot \boldsymbol{\alpha})) \equiv 4\mathbf{a} \cdot \mathbf{b} = 4(a_1b_1 + a_2b_2 + a_3b_3)$$

and

$$\text{tr}((a \cdot \alpha)(b \cdot \alpha)(c \cdot \alpha)(d \cdot \alpha)) \equiv 4(a \cdot b)(c \cdot d) - 4(a \cdot c)(b \cdot d) + 4(a \cdot d)(b \cdot c).$$

These identities are important in quantum electrodynamics.

Problem 6.   Given the orthonormal basis

$$x_1 = \begin{pmatrix} e^{i\phi} \cos\theta \\ \sin\theta \end{pmatrix}, \qquad x_2 = \begin{pmatrix} -\sin\theta \\ e^{-i\phi} \cos\theta \end{pmatrix}$$

in the vector space $\mathbf{C}^2$. Use this basis to find a basis in $\mathbf{C}^4$.

Solution 6.   A basis in $\mathbf{C}^4$ is given by

$$\{\, x_1 \otimes x_1, \quad x_1 \otimes x_2, \quad x_2 \otimes x_1, \quad x_2 \otimes x_2 \,\}$$

since

$$(x_j^* \otimes x_k^*)(x_m \otimes x_n) = \delta_{jm}\delta_{kn}$$

where $j, k, m, n = 1, 2$.

Problem 7.   Let A be an $m \times m$ matrix and B be an $n \times n$ matrix. The underlying field is $\mathbf{C}$. Let I_m, I_n be the $m \times m$ and $n \times n$ unit matrix, respectively.
(i) Show that
$$\text{tr}(A \otimes B) = \text{tr}(A)\text{tr}(B).$$

(ii) Show that

$$\text{tr}(A \otimes I_n + I_m \otimes B) = n\,\text{tr}(A) + m\,\text{tr}(B).$$

Solution 7.   (i) We have

$$\text{tr}(A \otimes B) = \sum_{j=1}^{m} \sum_{k=1}^{n} a_{jj} b_{kk}$$

$$= \left(\sum_{j=1}^{m} a_{jj} \right) \left(\sum_{k=1}^{n} b_{kk} \right)$$

$$= \text{tr}(A)\text{tr}(B).$$

(ii) Since the trace operation is linear and $\text{tr} I_n = n$ we find

$$\text{tr}(A \otimes I_n + I_m \otimes B) = \text{tr}(A \otimes I_n) + \text{tr}(I_m \otimes B) = n\,\text{tr}(A) + m\,\text{tr}(B).$$

Problem 8.   Let A be an arbitrary $n \times n$ matrix over $\mathbf{C}$. Show that

$$\exp(A \otimes I_n) \equiv \exp(A) \otimes I_n . \tag{1}$$

Solution 8.   Using the expansion of the exponential function

$$\exp(A \otimes I_n) = \sum_{k=0}^{\infty} \frac{(A \otimes I_n)^k}{k!}$$

$$= I_n \otimes I_n + \frac{1}{1!}(A \otimes I_n) + \frac{1}{2!}(A \otimes I_n)^2 + \frac{1}{3!}(A \otimes I_n)^3 + \cdots$$

and

$$(A \otimes I_n)^k = A^k \otimes I_n, \qquad k \in \mathbf{N}$$

we find identity (1).

Problem 9.   Let A, B be arbitrary $n \times n$ matrices over $\mathbf{C}$. Let I_n be the $n \times n$ unit matrix. Show that

$$\exp(A \otimes I_n + I_n \otimes B) \equiv \exp(A) \otimes \exp(B)$$

Solution 9.   The proof of this identity relies on

$$[A \otimes I_n, I_n \otimes B] = 0_{n^2}$$

where $[,]$ denotes the commutator, 0_{n^2} is the $n^2 \times n^2$ zero matrix and

$$(A \otimes I_n)^r (I_n \otimes B)^s \equiv (A^r \otimes I_n)(I_n \otimes B^s) \equiv A^r \otimes B^s, \qquad r, s \in \mathbf{N} .$$

Thus

$$\exp(A \otimes I_n + I_n \otimes B) = \sum_{j=0}^{\infty} \frac{(A \otimes I_n + I_n \otimes B)^j}{j!}$$

$$= \sum_{j=0}^{\infty} \sum_{k=0}^{j} \frac{1}{j!} \binom{j}{k} (A \otimes I_n)^k (I_n \otimes B)^{j-k}$$

$$= \sum_{j=0}^{\infty} \sum_{k=0}^{j} \frac{1}{j!} \binom{j}{k} (A^k \otimes B^{j-k})$$

$$= \left(\sum_{j=0}^{\infty} \frac{A^j}{j!} \right) \otimes \left(\sum_{k=0}^{\infty} \frac{B^k}{k!} \right)$$

$$= \exp(A) \otimes \exp(B) .$$

Problem 10.   Let A and B be arbitrary $n \times n$ matrices over $\mathbf{C}$. Prove or disprove the equation

$$e^{A \otimes B} = e^A \otimes e^B .$$

Solution 10.   Obviously this is not true in general. For example, let $A = B = I_n$. Then

$$e^{A \otimes B} = e^{I_{n^2}}$$

and

$$e^A \otimes e^B = e^{I_n} \otimes e^{I_n} \neq e^{I_{n^2}}.$$

Problem 11.   Let A be an $m \times m$ matrix and B be an $n \times n$ matrix. The underlying field is $\mathbf{C}$. The eigenvalues and eigenvectors of A are given by $\lambda_1, \lambda_2, \ldots, \lambda_m$ and $\mathbf{u}_1, \mathbf{u}_2, \ldots, \mathbf{u}_m$. The eigenvalues and eigenvectors of B are given by $\mu_1, \mu_2, \ldots, \mu_n$ and $\mathbf{v}_1, \mathbf{v}_2, \ldots, \mathbf{v}_n$. Let ϵ_1, ϵ_2 and ϵ_3 be real parameters. Find the eigenvalues and eigenvectors of the matrix

$$\epsilon_1 A \otimes B + \epsilon_2 A \otimes I_n + \epsilon_3 I_m \otimes B .$$

Solution 11.   Let $\mathbf{x} \in \mathbf{C}^m$ and $\mathbf{y} \in \mathbf{C}^n$. Then we have

$$(A \otimes B)(\mathbf{x} \otimes \mathbf{y}) = (A\mathbf{x}) \otimes (B\mathbf{y}),$$

$$(A \otimes I_n)(\mathbf{x} \otimes \mathbf{y}) = (A\mathbf{x}) \otimes \mathbf{y}, \qquad (I_m \otimes B)(\mathbf{x} \otimes \mathbf{y}) = \mathbf{x} \otimes (B\mathbf{y}) .$$

Thus the eigenvectors of the matrix $\epsilon_1 A \otimes B + \epsilon_2 A \otimes I_n + \epsilon_3 I_m \otimes B$ are

$$\mathbf{u}_j \otimes \mathbf{v}_k, \qquad j = 1, 2, \ldots, m \quad k = 1, 2, \ldots, n .$$

The corresponding eigenvalues are given by

$$\epsilon_1 \lambda_j \mu_k + \epsilon_2 \lambda_j + \epsilon_3 \mu_k .$$

Problem 12.   Let A, B be $n \times n$ matrices over $\mathbf{C}$. A *scalar product* can be defined as

$$\langle A, B \rangle := \operatorname{tr}(AB^*) .$$

The scalar product implies a *norm*

$$\|A\|^2 = \langle A, A \rangle = \operatorname{tr}(AA^*) .$$

This norm is called the *Hilbert-Schmidt norm*.

(i) Consider the *Dirac matrices*

$$\gamma_0 := \begin{pmatrix} 1 & 0 & 0 & 0 \\ 0 & 1 & 0 & 0 \\ 0 & 0 & -1 & 0 \\ 0 & 0 & 0 & -1 \end{pmatrix}, \qquad \gamma_1 := \begin{pmatrix} 0 & 0 & 0 & 1 \\ 0 & 0 & 1 & 0 \\ 0 & -1 & 0 & 0 \\ -1 & 0 & 0 & 0 \end{pmatrix}.$$

Calculate $\langle \gamma_0, \gamma_1 \rangle$.

(ii) Let U be a unitary $n \times n$ matrix. Find $\langle UA, UB \rangle$.

(iii) Let C, D be $m \times m$ matrices over **C**. Find $\langle A \otimes C, B \otimes D \rangle$.

Solution 12.   (i) We find

$$\langle \gamma_0, \gamma_1 \rangle = \operatorname{tr}(\gamma_0 \gamma_1^*) = 0.$$

(ii) Since

$$\operatorname{tr}(UA(UB)^*) = \operatorname{tr}(UAB^*U^*) = \operatorname{tr}(U^*UAB^*) = \operatorname{tr}(AB)$$

where we used the *cyclic invariance* for matrices, we find that

$$\langle UA, UB \rangle = \langle A, B \rangle.$$

Thus the scalar product is invariant under the unitary transformation.

(iii) Since

$$\begin{aligned} \operatorname{tr}((A \otimes C)(B \otimes D)^*) &= \operatorname{tr}((A \otimes C)(B^* \otimes D^*)) \\ &= \operatorname{tr}((AB^*) \otimes (CD^*)) \\ &= \operatorname{tr}(AB^*)\operatorname{tr}(CD^*) \end{aligned}$$

we find

$$\langle A \otimes C, B \otimes D \rangle = \langle A, B \rangle \langle C, D \rangle.$$

Problem 13.   Let T be the 4×4 matrix

$$T := \left(I_2 \otimes I_2 + \sum_{j=1}^{3} t_j \sigma_j \otimes \sigma_j \right)$$

where σ_j, $j = 1, 2, 3$ are the Pauli spin matrices and $-1 \le t_j \le +1$, $j = 1, 2, 3$. Find T^2.

Solution 13.   We have

$$T^2 = I_2 \otimes I_2 + 2 \sum_{j=1}^{3} t_j \sigma_j \otimes \sigma_j + \sum_{j=1}^{3} \sum_{k=1}^{3} t_j t_k (\sigma_j \sigma_k) \otimes (\sigma_j \sigma_k).$$

Since

$$\sigma_1\sigma_2 = i\sigma_3, \qquad \sigma_2\sigma_1 = -i\sigma_3$$

$$\sigma_2\sigma_3 = i\sigma_1, \qquad \sigma_3\sigma_2 = -i\sigma_1$$

$$\sigma_3\sigma_1 = i\sigma_2, \qquad \sigma_1\sigma_3 = -i\sigma_2$$

and $\sigma_1^2 = I_2$, $\sigma_2^2 = I_2$, $\sigma_3^2 = I_2$, we find

$$\sum_{j=1}^{3}\sum_{k=1}^{3} t_j t_k \sigma_j\sigma_k \otimes \sigma_j\sigma_k \equiv I_2 \otimes I_2 \sum_{j=1}^{3} t_j^2 - 2(t_1 t_2\sigma_3\otimes\sigma_3 + t_2 t_3\sigma_1\otimes\sigma_1 + t_3 t_1\sigma_2\otimes\sigma_2).$$

Therefore

$$T^2 = (I_2 \otimes I_2)\left(1 + \sum_{j=1}^{3} t_j^2\right)$$
$$+2(t_1 - t_2 t_3)\sigma_1 \otimes \sigma_1 + 2(t_2 - t_3 t_1)\sigma_2 \otimes \sigma_2 + 2(t_3 - t_1 t_2)\sigma_3 \otimes \sigma_3.$$

Problem 14.   Let U be a 2×2 unitary matrix and I_2 be the 2×2 identity matrix. Is the 4×4 matrix

$$V = \begin{pmatrix} 0 & 0 \\ 0 & 1 \end{pmatrix} \otimes U + \begin{pmatrix} e^{i\alpha} & 0 \\ 0 & 0 \end{pmatrix} \otimes I_2, \qquad \alpha \in \mathbf{R}$$

unitary?

Solution 14.   Since

$$\begin{pmatrix} 0 & 0 \\ 0 & 1 \end{pmatrix}\begin{pmatrix} e^{i\alpha} & 0 \\ 0 & 0 \end{pmatrix} = \begin{pmatrix} e^{i\alpha} & 0 \\ 0 & 0 \end{pmatrix}\begin{pmatrix} 0 & 0 \\ 0 & 1 \end{pmatrix} = \begin{pmatrix} 0 & 0 \\ 0 & 0 \end{pmatrix}$$

and $UU^* = I_2$ we obtain

$$VV^* = \begin{pmatrix} 0 & 0 \\ 0 & 1 \end{pmatrix} \otimes I_2 + \begin{pmatrix} 1 & 0 \\ 0 & 0 \end{pmatrix} \otimes I_2 = I_2 \otimes I_2.$$

Problem 15.   Let

$$\mathbf{x} = \begin{pmatrix} x_1 \\ x_2 \end{pmatrix}, \qquad x_1 x_1^* + x_2 x_2^* = 1$$

be an arbitrary normalized vector in $\mathbf{C}^2$. Can we construct a 4×4 unitary matrix U such that

$$U\left(\begin{pmatrix} x_1 \\ x_2 \end{pmatrix} \otimes \begin{pmatrix} 1 \\ 0 \end{pmatrix}\right) = \begin{pmatrix} x_1 \\ x_2 \end{pmatrix} \otimes \begin{pmatrix} x_1 \\ x_2 \end{pmatrix} ? \tag{1}$$

Prove or disprove this equation.

Solution 15. Such a matrix does not exist. This can be seen as follows. From the right-hand side of (1) we have

$$\begin{pmatrix} x_1 \\ x_2 \end{pmatrix} \otimes \begin{pmatrix} x_1 \\ x_2 \end{pmatrix} = \left(\begin{pmatrix} x_1 \\ 0 \end{pmatrix} + \begin{pmatrix} 0 \\ x_2 \end{pmatrix} \right) \otimes \left(\begin{pmatrix} x_1 \\ 0 \end{pmatrix} + \begin{pmatrix} 0 \\ x_2 \end{pmatrix} \right)$$

$$= \begin{pmatrix} x_1 \\ 0 \end{pmatrix} \otimes \begin{pmatrix} x_1 \\ 0 \end{pmatrix} + \begin{pmatrix} x_1 \\ 0 \end{pmatrix} \otimes \begin{pmatrix} 0 \\ x_2 \end{pmatrix} + \begin{pmatrix} 0 \\ x_2 \end{pmatrix} \otimes \begin{pmatrix} x_1 \\ 0 \end{pmatrix} + \begin{pmatrix} 0 \\ x_2 \end{pmatrix} \otimes \begin{pmatrix} 0 \\ x_2 \end{pmatrix}.$$

However from the left-hand side of (1) we find

$$U \left(\begin{pmatrix} x_1 \\ x_2 \end{pmatrix} \otimes \begin{pmatrix} 1 \\ 0 \end{pmatrix} \right) = U \left(\begin{pmatrix} x_1 \\ 0 \end{pmatrix} \otimes \begin{pmatrix} 1 \\ 0 \end{pmatrix} + \begin{pmatrix} 0 \\ x_2 \end{pmatrix} \otimes \begin{pmatrix} 1 \\ 0 \end{pmatrix} \right)$$

$$= \begin{pmatrix} x_1 \\ 0 \end{pmatrix} \otimes \begin{pmatrix} x_1 \\ 0 \end{pmatrix} + \begin{pmatrix} 0 \\ x_2 \end{pmatrix} \otimes \begin{pmatrix} 0 \\ x_2 \end{pmatrix}$$

where we used the linearity of U. Comparing these two equations we find a contradiction. This is the *no cloning theorem*. However equation (1) does hold when the following equation is satisfied

$$\begin{pmatrix} x_1 \\ 0 \end{pmatrix} \otimes \begin{pmatrix} 0 \\ x_2 \end{pmatrix} + \begin{pmatrix} 0 \\ x_2 \end{pmatrix} \otimes \begin{pmatrix} x_1 \\ 0 \end{pmatrix} = \begin{pmatrix} 0 \\ 0 \end{pmatrix} \otimes \begin{pmatrix} 0 \\ 0 \end{pmatrix}.$$

Therefore $x_1 x_2 = 0$. Thus at least one of x_1 and x_2 must be zero. It is still possible to clone elements of a known orthonormal basis.

Problem 16. Let A_j $(j = 1, 2, \ldots, k)$ be matrices of size $m_j \times n_j$. We introduce the notation

$$\otimes_{j=1}^{k} A_j = (\otimes_{j=1}^{k-1} A_j) \otimes A_k = A_1 \otimes A_2 \otimes \cdots \otimes A_k.$$

Consider the binary matrices

$$J_{00} = \begin{pmatrix} 1 & 0 \\ 0 & 0 \end{pmatrix}, \quad J_{10} = \begin{pmatrix} 0 & 0 \\ 1 & 0 \end{pmatrix}, \quad J_{01} = \begin{pmatrix} 0 & 1 \\ 0 & 0 \end{pmatrix}, \quad J_{11} = \begin{pmatrix} 0 & 0 \\ 0 & 1 \end{pmatrix}.$$

(i) Calculate

$$\otimes_{j=1}^{n} (J_{00} + J_{01} + J_{11})$$

for $k = 1$, $k = 2$, $k = 3$ and $k = 8$. Give an interpretation of the result when each entry in the matrix represents a pixel (1 for black and 0 for white). This means we use the Kronecker product for representing images.

(ii) Calculate

$$(\otimes_{j=1}^{k} (J_{00} + J_{01} + J_{10} + J_{11})) \otimes \begin{pmatrix} 0 & 1 \\ 1 & 0 \end{pmatrix}$$

for $k = 2$ and give an interpretation as an image, i.e., each entry 0 is identified with a black pixel and an entry 1 with a white pixel. Discuss the case for arbitrary k.

Solution 16. (i) We obtain

$$J_{00} + J_{01} + J_{11} = \begin{pmatrix} 1 & 1 \\ 0 & 1 \end{pmatrix}$$

$$(J_{00} + J_{01} + J_{11}) \otimes (J_{00} + J_{01} + J_{11}) = \begin{pmatrix} 1 & 1 & 1 & 1 \\ 0 & 1 & 0 & 1 \\ 0 & 0 & 1 & 1 \\ 0 & 0 & 0 & 1 \end{pmatrix}.$$

For $k = 3$ we obtain the 8×8 matrix

$$\begin{pmatrix} 1 & 1 & 1 & 1 & 1 & 1 & 1 & 1 \\ 0 & 1 & 0 & 1 & 0 & 1 & 0 & 1 \\ 0 & 0 & 1 & 1 & 0 & 0 & 1 & 1 \\ 0 & 0 & 0 & 1 & 0 & 0 & 0 & 1 \\ 0 & 0 & 0 & 0 & 1 & 1 & 1 & 1 \\ 0 & 0 & 0 & 0 & 0 & 1 & 0 & 1 \\ 0 & 0 & 0 & 0 & 0 & 0 & 1 & 1 \\ 0 & 0 & 0 & 0 & 0 & 0 & 0 & 1 \end{pmatrix}.$$

For n larger and larger the image approaches the *Sierpinski triangle*. For $k = 8$ we have an 256×256 matrix which provides a good approximation for the Sierpinski triangle.

(ii) Note that

$$J_{00} + J_{01} + J_{10} + J_{11} = \begin{pmatrix} 1 & 1 \\ 1 & 1 \end{pmatrix}.$$

Then for $n = 2$ we obtain the 8×8 matrix

$$\begin{pmatrix} 0 & 1 & 0 & 1 & 0 & 1 & 0 & 1 \\ 1 & 0 & 1 & 0 & 1 & 0 & 1 & 0 \\ 0 & 1 & 0 & 1 & 0 & 1 & 0 & 1 \\ 1 & 0 & 1 & 0 & 1 & 0 & 1 & 0 \\ 0 & 1 & 0 & 1 & 0 & 1 & 0 & 1 \\ 1 & 0 & 1 & 0 & 1 & 0 & 1 & 0 \\ 0 & 1 & 0 & 1 & 0 & 1 & 0 & 1 \\ 1 & 0 & 1 & 0 & 1 & 0 & 1 & 0 \end{pmatrix}$$

which is check-board. For all n we find a check-board pattern.

Problem 17. Consider the Pauli spin matrices $\sigma = (\sigma_1, \sigma_2, \sigma_3)$. Let $\mathbf{q}$, $\mathbf{r}$, $\mathbf{s}$, $\mathbf{t}$ be unit vectors in $\mathbf{R}^3$. We define

$$Q := \mathbf{q} \cdot \sigma, \quad R := \mathbf{r} \cdot \sigma, \quad S := \mathbf{s} \cdot \sigma, \quad T := \mathbf{t} \cdot \sigma$$

where $\mathbf{q} \cdot \sigma := q_1\sigma_1 + q_2\sigma_2 + q_3\sigma_3$. Calculate

$$(Q \otimes S + R \otimes S + R \otimes T - Q \otimes T)^2.$$

Express the result using commutators.

Solution 17. Using that for $j, k = 1, 2, 3$ we have

$$\sigma_j\sigma_k = \delta_{jk}I_2 + i\sum_{\ell=1}^{3}\epsilon_{jk\ell}\sigma_\ell$$

with $\epsilon_{123} = \epsilon_{231} = \epsilon_{312} = 1$, $\epsilon_{321} = \epsilon_{213} = \epsilon_{132} = -1$ and 0 otherwise we obtain

$$(Q \otimes S + R \otimes S + R \otimes T - Q \otimes T)^2 = 4I_2 \otimes I_2 + [Q, R] \otimes [S, T].$$

Problem 18. Let A and X be $n \times n$ matrices over $\mathbf{C}$. Assume that

$$[X, A] = 0_n.$$

Calculate $[X \otimes I_n + I_n \otimes X, A \otimes A]$.

Solution 18. We have

$$\begin{aligned}
[X \otimes I_n + I_n \otimes X, A \otimes A] &= (X \otimes I_n + I_n \otimes X)(A \otimes A) \\
&\quad -(A \otimes A)(X \otimes I_n + I_n \otimes X) \\
&= (X \otimes I_n)(A \otimes A) + (I_n \otimes X)(A \otimes A) \\
&\quad -(A \otimes A)(X \otimes I_n) - (A \otimes A)(I_n \otimes X) \\
&= (XA) \otimes A + A \otimes (XA) \\
&\quad -(AX) \otimes A - A \otimes (AX).
\end{aligned}$$

Since $AX = XA$ we obtain

$$[X \otimes I_n + I_n \otimes X, A \otimes A] = 0_{n^2}$$

where 0_{n^2} is the $n^2 \times n^2$ zero matrix.

Problem 19. A square matrix is called a *stochastic matrix* if each entry is nonnegative and the sum of the entries in each row is 1. Let A, B be $n \times n$ stochastic matrices. Is $A \otimes B$ a stochastic matrix?

Solution 19. The answer is yes. For example the sum of the first row of the matrix $A \otimes B$ is

$$a_{11}\sum_{j=1}^{n}b_{1j} + a_{12}\sum_{j=1}^{n}b_{1j} + \cdots + a_{1n}\sum_{j=1}^{n}b_{1j} = \sum_{j=1}^{n}a_{1j} = 1.$$

Analogously for the other rows.

Problem 20.   Let X be an $m \times m$ and B be an $n \times n$ matrix. The *direct sum* is the $(m + n) \times (m + n)$ matrix

$$X \oplus Y = \begin{pmatrix} A & 0 \\ 0 & B \end{pmatrix}.$$

Let A be an $n \times n$ matrix, B be an $m \times m$ matrix and C be an $p \times p$ matrix. Then we have the identity

$$(A \oplus B) \otimes C \equiv (A \otimes C) \oplus (B \otimes C).$$

Is

$$A \otimes (B \oplus C) = (A \otimes B) \oplus (A \otimes C)$$

true?

Solution 20.   This is not true in general. For example, let

$$A = \begin{pmatrix} 2 & 1 \\ 0 & 3 \end{pmatrix}, \quad B = \begin{pmatrix} 1 & 1 \\ 1 & 1 \end{pmatrix}, \quad C = \begin{pmatrix} 0 & 1 \\ 1 & 0 \end{pmatrix}.$$

Then $A \otimes (B \oplus C) \neq (A \otimes B) \oplus (A \otimes C)$.

Problem 21.   Let A, B be 2×2 matrices, C a 3×3 matrix and D a 1×1 matrix. Find the condition on these matrices such that

$$A \otimes B = C \oplus D$$

where $\oplus$ denotes the *direct sum*. We assume that D is nonzero.

Solution 21.   Since

$$A \otimes B = \begin{pmatrix} a_{11}b_{11} & a_{11}b_{12} & a_{12}b_{11} & a_{12}b_{12} \\ a_{11}b_{21} & a_{11}b_{22} & a_{12}b_{21} & a_{12}b_{22} \\ a_{21}b_{11} & a_{21}b_{12} & a_{22}b_{11} & a_{22}b_{12} \\ a_{21}b_{21} & a_{21}b_{22} & a_{22}b_{21} & a_{22}b_{22} \end{pmatrix}$$

and

$$C \oplus D = \begin{pmatrix} c_{11} & c_{12} & c_{13} & 0 \\ c_{21} & c_{22} & c_{23} & 0 \\ c_{31} & c_{32} & c_{33} & 0 \\ 0 & 0 & 0 & d \end{pmatrix}$$

we obtain from the last row and column of the two matrices $A \otimes B$

$$a_{12}b_{12} = a_{12}b_{22} = a_{22}b_{12} = 0$$

$$a_{21}b_{21} = a_{21}b_{22} = a_{22}b_{21} = 0$$

and $a_{22}b_{22} = d$. Since $d \neq 0$ we have $a_{22} \neq 0$ and $b_{22} \neq 0$. Thus it follows that

$$a_{12} = a_{21} = b_{12} = b_{21} = 0.$$

Therefore

$$c_{12} = c_{21} = c_{23} = c_{32} = c_{13} = c_{31} = 0$$

and

$$c_{11} = a_{11}b_{11}, \quad c_{22} = a_{11}b_{22}, \quad c_{33} = a_{22}b_{11}, \quad d = a_{22}b_{22}.$$

Problem 22. With each $m \times n$ matrix Y we associate the column vector vec$Y$ of length $m \times n$ defined by

$$\text{vec} Y := (y_{11}, \ldots, y_{m1}, y_{12}, \ldots, y_{m2}, \ldots, y_{1n}, \ldots, y_{mn})^T.$$

Let A be an $m \times n$ matrix, B an $p \times q$ matrix, and C an $m \times q$ matrix. Let X be an unknown $n \times p$ matrix. Show that the matrix equation

$$AXB = C$$

is equivalent to the system of qm equations in np unknowns given by

$$(B^T \otimes A)\text{vec} X = \text{vec} C.$$

that is, $\text{vec}(AXB) = (B^T \otimes A)\text{vec} X$.

Solution 22. For a given matrix Y, let Y_k denote the kth column of Y. Let $B = (b_{ij})$. Then

$$(AXB)_k = A(XB)_k = AXB_k$$
$$= A \left(\sum_{i=1}^{p} b_{ik} X_i \right) = (b_{1k}A \quad b_{2k}A \quad \ldots \quad b_{pk}A)\text{vec} X$$
$$= (B_k^T \otimes A)\text{vec} X.$$

Thus

$$\text{vec}(AXB) = \begin{pmatrix} B_1^T \otimes A \\ \vdots \\ B_q^T \otimes A \end{pmatrix} \text{vec} X$$
$$= (B^T \otimes A)\text{vec} X$$

since the transpose of a column of B is a row of B^T. Thus,

$$\text{vec} C = \text{vec}(AXB) = (B^T \otimes A)\text{vec} X.$$

Problem 23. Let A, B, D be $n \times n$ matrices and I_n the $n \times n$ identity matrix. Use the result from the problem above to prove that

$$AX + XB = D$$

can be written as

$$((I_n \otimes A) + (B^T \otimes I_n))\text{vec}X = \text{vec}D. \tag{1}$$

Solution 23. From the result of the problem given above and that the vec operation is linear we have

$$
\begin{aligned}
\text{vec}(AX + XB) &= \text{vec}(AX) + \text{vec}(XB) \\
&= \text{vec}(AXI_n) + \text{vec}(I_nXB) \\
&= (I_n \otimes A)\text{vec}X + (B^T \otimes I_n)\text{vec}X \\
&= ((I_n \otimes A) + (B^T \otimes I_n))\text{vec}X \\
&= \text{vec}D.
\end{aligned}
$$

Problem 24. Let A be an $n \times n$ matrix and I_m be the $m \times m$ identity matrix. Show that

$$\sin(A \otimes I_m) \equiv \sin(A) \otimes I_m. \tag{1}$$

Solution 24. Let $k \in \mathbf{N}$. Since

$$(A \otimes I_m)^k = A^k \otimes I_m$$

we find identity (1).

Problem 25. Let A be an $n \times n$ matrix and B be an $m \times m$ matrix. Is

$$\sin(A \otimes I_m + I_n \otimes B) \equiv (\sin(A)) \otimes (\cos(B)) + (\cos(A)) \otimes (\sin(B))? \tag{1}$$

Prove or disprove.

Solution 25. Since

$$[A \otimes I_m, I_n \otimes B] = 0_{n \times m}$$

and using the result from the problem above

$$\sin(A \otimes I_m) = \sin(A) \otimes I_m, \qquad \cos(I_n \otimes B) = I_n \otimes \cos(B)$$

we find identity (1).

Problem 26. Let A, B be $m \times n$ matrices. The *Hadamard product* of A and B is defined by the $m \times n$ matrix

$$A \bullet B := (a_{ij}b_{ij}).$$

We consider the case $m = n$. There exists an $n^2 \times n$ *selection matrix* J such that

$$A \bullet B = J^T(A \otimes B)J$$

where J^T is defined as the $n \times n^2$ matrix

$$[E_{11} \; E_{22} \; \ldots \; E_{nn}]$$

with E_{ii} the $n \times n$ matrix of zeros except for a 1 in the (i, i)th position. Prove this identity for the special case $n = 2$.

Solution 26. From the definition for the selection matrix we have

$$J^T = \begin{pmatrix} 1 & 0 & 0 & 0 \\ 0 & 0 & 0 & 1 \end{pmatrix}$$

and therefore

$$J = \begin{pmatrix} 1 & 0 \\ 0 & 0 \\ 0 & 0 \\ 0 & 1 \end{pmatrix}.$$

We have

$$A \bullet B = \begin{pmatrix} a_{11}b_{11} & a_{12}b_{12} \\ a_{21}b_{21} & a_{22}b_{22} \end{pmatrix}.$$

Now

$$J^T(A \otimes B)J = \begin{pmatrix} 1 & 0 & 0 & 0 \\ 0 & 0 & 0 & 1 \end{pmatrix} \begin{pmatrix} a_{11}b_{11} & a_{11}b_{12} & a_{12}b_{11} & a_{12}b_{12} \\ a_{11}b_{21} & a_{11}b_{22} & a_{12}b_{21} & a_{12}b_{22} \\ a_{21}b_{11} & a_{21}b_{12} & a_{22}b_{11} & a_{22}b_{12} \\ a_{21}b_{21} & a_{21}b_{22} & a_{22}b_{21} & a_{22}b_{22} \end{pmatrix} \begin{pmatrix} 1 & 0 \\ 0 & 0 \\ 0 & 0 \\ 0 & 1 \end{pmatrix}$$

$$= \begin{pmatrix} 1 & 0 & 0 & 0 \\ 0 & 0 & 0 & 1 \end{pmatrix} \begin{pmatrix} a_{11}b_{11} & a_{12}b_{12} \\ a_{11}b_{21} & a_{12}b_{22} \\ a_{21}b_{11} & a_{22}b_{12} \\ a_{21}b_{21} & a_{22}b_{22} \end{pmatrix}$$

$$= \begin{pmatrix} a_{11}b_{11} & a_{12}b_{12} \\ a_{21}b_{21} & a_{22}b_{22} \end{pmatrix}.$$

Problem 27. Let σ_x, σ_y, σ_z be the *Pauli spin matrices*

$$\sigma_x := \begin{pmatrix} 0 & 1 \\ 1 & 0 \end{pmatrix}, \qquad \sigma_y := \begin{pmatrix} 0 & -i \\ i & 0 \end{pmatrix}, \qquad \sigma_z := \begin{pmatrix} 1 & 0 \\ 0 & -1 \end{pmatrix}.$$

(i) Find

$$R_{1x}(\alpha) := \exp(-i\alpha(\sigma_x \otimes I_2)), \qquad R_{1y}(\alpha) := \exp(-i\alpha(\sigma_y \otimes I_2))$$

where $\alpha \in \mathbf{R}$ and I_2 denotes the 2×2 unit matrix.
(ii) Consider the special case $R_{1x}(\alpha = \pi/2)$ and $R_{1y}(\alpha = \pi/4)$. Calculate $R_{1x}(\pi/2)R_{1y}(\pi/4)$. Discuss.

Solution 27.   (i) We have

$$\exp(-i\alpha(\sigma_x \otimes I_2)) := \sum_{k=0}^{\infty} \frac{(-i\alpha(\sigma_x \otimes I_2))^k}{k!}.$$

Since $\sigma_x^2 = I_2$ we have

$$(\sigma_x \otimes I_2)^2 = I_2 \otimes I_2.$$

Thus we find

$$\exp(-i\alpha(\sigma_x \otimes I_2)) = (I_2 \otimes I_2)\cos\alpha + e^{-i\pi/2}(\sigma_x \otimes I_2)\sin\alpha$$

where we used $\exp(-i\pi/2) = -i$. Analogously, we find

$$\exp(-i\alpha(\sigma_y \otimes I_2)) = (I_2 \otimes I_2)\cos\alpha + e^{-i\pi/2}(\sigma_y \otimes I_2)\sin\alpha$$

since

$$(\sigma_y \otimes I_2)^2 = I_2 \otimes I_2.$$

(ii) Since $\sin(\pi/2) = 1$, $\cos(\pi/2) = 0$ we arrive at

$$R_{1x}(\pi/2) = e^{-i\pi/2}(\sigma_x \otimes I_2).$$

From $\sin(\pi/4) = \sqrt{2}/2$, $\cos(\pi/4) = \sqrt{2}/2$ it follows that

$$R_{1y}(\pi/4) = \frac{1}{\sqrt{2}}(I_2 \otimes I_2) + \frac{1}{\sqrt{2}}e^{-i\pi/2}(\sigma_y \otimes I_2).$$

Thus

$$R_{1x}(\pi/2)R_{1y}(\pi/4) = \frac{e^{-i\pi/2}}{\sqrt{2}}(\sigma_x \otimes I_2) + \frac{e^{-i\pi/2}}{\sqrt{2}}(\sigma_z \otimes I_2)$$

where we used that $\sigma_x\sigma_y = i\sigma_z$. Therefore

$$R_{1x}(\pi/2)R_{1y}(\pi/4) = \frac{e^{-i\pi/2}}{\sqrt{2}}(\sigma_x + \sigma_z) \otimes I_2$$

where

$$\frac{1}{\sqrt{2}}(\sigma_x + \sigma_z) = \frac{1}{\sqrt{2}}\begin{pmatrix} 1 & 1 \\ 1 & -1 \end{pmatrix}$$

is the *Hadamard matrix*. All the single operations are in the Lie group $SU(2)$ whose determinant is $+1$, while the determinant of the Hadamard matrix is -1. Thus the overall phase is unavoidable.

Problem 28. Let $\mathbf{x}, \mathbf{y} \in \mathbf{R}^2$. Find a 4×4 matrix A (*flip operator*) such that
$$A(\mathbf{x} \otimes \mathbf{y}) = \mathbf{y} \otimes \mathbf{x}.$$

Solution 28. Since
$$\mathbf{x} \otimes \mathbf{y} = \begin{pmatrix} x_1 y_1 \\ x_1 y_2 \\ x_2 y_1 \\ x_2 y_2 \end{pmatrix}, \qquad \mathbf{y} \otimes \mathbf{x} = \begin{pmatrix} y_1 x_1 \\ y_1 x_2 \\ y_2 x_1 \\ y_2 x_2 \end{pmatrix}$$
we find the permutation matrix
$$A = \begin{pmatrix} 1 & 0 & 0 & 0 \\ 0 & 0 & 1 & 0 \\ 0 & 1 & 0 & 0 \\ 0 & 0 & 0 & 1 \end{pmatrix}.$$

Problem 29. Let σ_x, σ_y and σ_z be the Pauli spin matrices
$$\sigma_x := \begin{pmatrix} 0 & 1 \\ 1 & 0 \end{pmatrix}, \qquad \sigma_y := \begin{pmatrix} 0 & -i \\ i & 0 \end{pmatrix}, \qquad \sigma_z := \begin{pmatrix} 1 & 0 \\ 0 & -1 \end{pmatrix}.$$
We define $\sigma_+ := \sigma_x + i\sigma_y$ and $\sigma_- := \sigma_x - i\sigma_y$. Let
$$c_k^* := \sigma_z \otimes \sigma_z \otimes \cdots \otimes \sigma_z \otimes \left(\frac{1}{2}\sigma_+\right) \otimes I_2 \otimes I_2 \otimes \cdots \otimes I_2$$
where σ_+ is on the kth position and we have $N - 1$ Kronecker products. Thus c_k^* is a $2^N \times 2^N$ matrix.
(i) Find c_k.
(ii) Find the anticommutators $[c_k, c_j]_+$ and $[c_k^*, c_j]_+$.
(iii) Find $c_k c_k$ and $c_k^* c_k^*$.

Solution 29. (i) Since $\sigma_+^* = \sigma_-$ we obtain
$$c_k := \sigma_z \otimes \sigma_z \otimes \cdots \otimes \sigma_z \otimes \left(\frac{1}{2}\sigma_-\right) \otimes I_2 \otimes I_2 \otimes \cdots \otimes I_2$$

(ii) Using $(A \otimes B)(C \otimes D) = (AC) \otimes (BD)$ we find that
$$[c_k, c_j]_+ := 0_{2^N}, \qquad [c_k^*, c_j]_+ := \delta_{kj} I_{2^N}$$

where δ_{jk} is the Kronecker delta with 1 for $j = k$ and 0 otherwise, I_{2^N} is the $2^N \times 2^N$ identity matrix and 0_{2^N} is the $2^N \times 2^N$ zero matrix.
(iii) Since $\sigma_+\sigma_+ = 0_2$ and $\sigma_-\sigma_- = 0_2$ we have

$$c_k c_k = 0_{2^N}, \qquad c_k^* c_k^* = 0_{2^N}.$$

This is the *Pauli exclusion principle*.

Problem 30. Using the definitions from the previous problem we define

$$s_{-,j} := \frac{1}{2}(\sigma_{x,j} - i\sigma_{y,j}) = \frac{1}{2}\sigma_{-,j}, \qquad s_{+,j} := \frac{1}{2}(\sigma_{x,j} + i\sigma_{y,j}) = \frac{1}{2}\sigma_{+,j}$$

and

$$c_1 = s_{-,1}$$

$$c_j = \exp\left(i\pi \sum_{\ell=1}^{j-1} s_{+,\ell} s_{-,\ell}\right) s_{-,j} \quad \text{for} \quad j = 2, 3, \ldots$$

(i) Find c_j^*.
(ii) Find the inverse transformation.
(iii) Calculate $c_j^* c_j$.

Solution 30. (i) Obviously we find

$$c_1^* = s_{+,1}$$

$$c_j^* = \exp\left(i\pi \sum_{\ell=1}^{j-1} s_{+,\ell} s_{-,\ell}\right) s_{+,j} \quad \text{for} \quad j = 2, 3, \ldots$$

(ii) The inverse transformation is given by

$$s_{1,-} = c_1$$

$$s_{j,-} = \exp\left(i\pi \sum_{\ell=1}^{j-1} c_\ell^* c_\ell\right) c_j \quad \text{for} \quad j = 2, 3, \ldots$$

$$s_{1,+} = c_1^*$$

$$s_{j,+} = \exp\left(i\pi \sum_{\ell=1}^{j-1} c_\ell^* c_\ell\right) c_j^* \quad \text{for} \quad j = 2, 3, \ldots$$

(iii) Straightforward calculation yields

$$c_j^* c_j = s_{+,j} s_{-,j}.$$

Problem 31. Let A, B, C, D be symmetric $n \times n$ matrices over **R**. Assume that these matrices commute with each other. Consider the $4n \times 4n$ matrix

$$H = \begin{pmatrix} A & B & C & D \\ -B & A & D & -C \\ -C & -D & A & B \\ -D & C & -B & A \end{pmatrix}.$$

(i) Calculate HH^T and express the result using the Kronecker product.
(ii) Assume that $A^2 + B^2 + C^2 + D^2 = 4nI_n$.

Solution 31. (i) We find

$$H^T H = (A^2 + B^2 + C^2 + D^2) \otimes I_4 .$$

(ii) For the special case with $A^2 + B^2 + C^2 + D^2 = 4nI_n$ we find that H is a Hadamard matrix, since

$$HH^T = 4nI_n \otimes I_4 = 4nI_{4n} .$$

This is the *Williamson construction* of Hadamard matrices.

Problem 32. Can the 4×4 matrix

$$C = \begin{pmatrix} 1 & 0 & 0 & 1 \\ 0 & 1 & 1 & 0 \\ 0 & 1 & -1 & 0 \\ 1 & 0 & 0 & -1 \end{pmatrix}$$

be written as the Kronecker product of two 2×2 matrices A and B, i.e. $C = A \otimes B$?

Solution 32. From $C = A \otimes B$ we find the conditions

$$a_{11}b_{11} = c_{11} \quad a_{11}b_{12} = c_{12} \quad a_{11}b_{21} = c_{21} \quad a_{11}b_{22} = c_{22}$$
$$a_{12}b_{11} = c_{13} \quad a_{12}b_{12} = c_{14} \quad a_{12}b_{21} = c_{23} \quad a_{12}b_{22} = c_{24}$$
$$a_{21}b_{11} = c_{31} \quad a_{21}b_{12} = c_{32} \quad a_{21}b_{21} = c_{41} \quad a_{21}b_{22} = c_{42}$$
$$a_{22}b_{11} = c_{33} \quad a_{22}b_{12} = c_{34} \quad a_{22}b_{21} = c_{43} \quad a_{22}b_{22} = c_{44} .$$

Since

$$c_{12} = c_{13} = c_{21} = c_{24} = c_{31} = c_{34} = c_{42} = c_{43} = 0$$

and

$$c_{11} = c_{14} = c_{22} = c_{23} = c_{32} = c_{41} = 1, \qquad c_{33} = c_{44} = -1$$

we obtain

$$a_{11}b_{11} = 1 \quad a_{11}b_{12} = 0 \quad a_{11}b_{21} = 0 \quad a_{11}b_{22} = 1$$
$$a_{12}b_{11} = 0 \quad a_{12}b_{12} = 1 \quad a_{12}b_{21} = 1 \quad a_{12}b_{22} = 0$$
$$a_{21}b_{11} = 0 \quad a_{21}b_{12} = 1 \quad a_{21}b_{21} = 1 \quad a_{21}b_{22} = 0$$
$$a_{22}b_{11} = -1 \quad a_{22}b_{12} = 0 \quad a_{22}b_{21} = 0 \quad a_{22}b_{22} = -1.$$

Thus from $a_{11}b_{11} = 1$ we obtain that $a_{11} \neq 0$. Then from $a_{11}b_{12} = 0$ we obtain $b_{12} = 0$. However $a_{21}b_{12} = 1$. This is a contradiction. Thus C cannot be written as a Kronecker product of 2×2 matrices.

Problem 33.   Let A, B, C be $n \times n$ matrices. Assume that

$$[A, B] = 0_n, \qquad [A, C] = 0_n.$$

Let

$$X := I_n \otimes A + A \otimes I_n, \qquad Y := I_n \otimes B + B \otimes I_n + A \otimes C.$$

Calculate the commutator $[X, Y]$.

Solution 33.   We have

$$
\begin{aligned}
[X, Y] &= (I_n \otimes A + A \otimes I_n)(I_n \otimes B + B \otimes I_n + A \otimes C) \\
&\quad -(I_n \otimes B + B \otimes I_n + A \otimes C)(I_n \otimes A + A \otimes I_n) \\
&= I_n \otimes (AB) + A \otimes B + B \otimes A + (AB) \otimes I_n + A \otimes (AC) + A^2 \otimes C \\
&\quad -I_n \otimes (BA) - A \otimes B - B \otimes A - (BA) \otimes I_n - A \otimes (CA) - A^2 \otimes C \\
&= I_n \otimes (AB - BA) + (AB - BA) \otimes I_n + A \otimes (AC - CA) \\
&= I_n \otimes [A, B] + [A, B] \otimes I_n + A \otimes [A, C] \\
&= 0_{n^2}.
\end{aligned}
$$

Problem 34.   Let $\mathbf{x}, \mathbf{y}, \mathbf{z} \in \mathbf{R}^n$. We define a *wedge product*

$$\mathbf{x} \wedge \mathbf{y} := \mathbf{x} \otimes \mathbf{y} - \mathbf{y} \otimes \mathbf{x}.$$

Show that

$$(\mathbf{x} \wedge \mathbf{y}) \wedge \mathbf{z} + (\mathbf{z} \wedge \mathbf{x}) \wedge \mathbf{y} + (\mathbf{y} \wedge \mathbf{z}) \wedge \mathbf{x} = \mathbf{0}.$$

Solution 34.   We have

$$(\mathbf{x} \wedge \mathbf{y}) \wedge \mathbf{z} = \mathbf{x} \otimes \mathbf{y} \otimes \mathbf{z} - \mathbf{y} \otimes \mathbf{x} \otimes \mathbf{z} - \mathbf{z} \otimes \mathbf{x} \otimes \mathbf{y} + \mathbf{z} \otimes \mathbf{y} \otimes \mathbf{x}$$

$$(\mathbf{z} \wedge \mathbf{x}) \wedge \mathbf{y} = \mathbf{z} \otimes \mathbf{x} \otimes \mathbf{y} - \mathbf{x} \otimes \mathbf{z} \otimes \mathbf{y} - \mathbf{y} \otimes \mathbf{z} \otimes \mathbf{x} + \mathbf{y} \otimes \mathbf{x} \otimes \mathbf{z}$$

$$(\mathbf{y} \wedge \mathbf{z}) \wedge \mathbf{x} = \mathbf{y} \otimes \mathbf{z} \otimes \mathbf{x} - \mathbf{z} \otimes \mathbf{y} \otimes \mathbf{x} - \mathbf{x} \otimes \mathbf{y} \otimes \mathbf{z} + \mathbf{x} \otimes \mathbf{z} \otimes \mathbf{y}.$$

Thus equation (1) follows.

Problem 35. Let V and W be the unitary matrices

$$V = \exp(i(\pi/4)\sigma_x) \otimes \exp(i(\pi/4)\sigma_x)$$
$$W = \exp(i(\pi/4)\sigma_y) \otimes \exp(i(\pi/4)\sigma_y).$$

Calculate

$$V^*(\sigma_z \otimes \sigma_z)V, \qquad W^*(\sigma_z \otimes \sigma_z)W.$$

Solution 35. Using

$$\exp(-i(\pi/4)\sigma_x)\sigma_z \exp(i(\pi/4)\sigma_x) = -\sigma_y$$

we obtain

$$V^*(\sigma_z \otimes \sigma_z)V = (e^{-i(\pi/4)\sigma_x} \otimes e^{-i(\pi/4)\sigma_x})(\sigma_z \otimes \sigma_z)(e^{i(\pi/4)\sigma_x} \otimes e^{i(\pi/4)\sigma_x})$$
$$= (e^{-i(\pi/4)\sigma_x}\sigma_z e^{i(\pi/4)\sigma_x}) \otimes (e^{-i(\pi/4)\sigma_x}\sigma_z e^{i(\pi/4)\sigma_x})$$
$$= \sigma_y \otimes \sigma_y.$$

Analogously using

$$\exp(-i(\pi/4)\sigma_y)\sigma_z \exp(i(\pi/4)\sigma_y) = \sigma_x$$

we obtain

$$W^*(\sigma_z \otimes \sigma_z)W = \sigma_x \otimes \sigma_x.$$

Chapter 10

Norms and Scalar Products

A linear space V is called a *normed space*, if for every $\mathbf{v} \in V$ there is associated a real number $\|\mathbf{v}\|$, the norm of the vector $\mathbf{v}$, such that

$$\|\mathbf{v}\| \geq 0, \qquad \|\mathbf{v}\| = 0 \text{ iff } \mathbf{v} = 0$$
$$\|c\mathbf{v}\| = |c|\,\|\mathbf{v}\| \quad \text{where } c \in \mathbf{C}$$
$$\|\mathbf{v} + \mathbf{w}\| \leq \|\mathbf{v}\| + \|\mathbf{w}\| \quad \text{for all } \mathbf{v}, \mathbf{w} \in V.$$

Consider $\mathbf{C}^n$ and $\mathbf{v} = (v_1, v_2, \ldots, v_n)$. The most common *vector norms* are:

(i) The *Euclidean norm*: $\|\mathbf{v}\|_2 := \sqrt{\mathbf{v}^*\mathbf{v}}$
(ii) The ℓ_1 norm: $\|\mathbf{v}\|_1 := |v_1| + |v_2| + \cdots + |v_n|$
(iii) The ℓ_∞ norm: $\|\mathbf{v}\|_\infty := \max(|v_1|, |v_2|, \ldots, |v_n|)$
(iv) The ℓ_p norm $(p \geq 1)$: $\|\mathbf{v}\|_p := (|v_1|^p + |v_2|^p + \cdots + |v_n|^p)^{1/p}$

Let A be an $n \times n$ matrix over $\mathbf{C}$ and $\mathbf{x} \in \mathbf{C}^n$. Each vector norm induces the *matrix norm*

$$\|A\| := \max_{\|\mathbf{x}\|=1} \|A\mathbf{x}\|.$$

Another important matrix norm is

$$\|A\| := \sqrt{\operatorname{tr}(AA^*)}.$$

This is called the *Frobenius norm*.

Problem 1. Consider the vector $(\mathbf{v} \in \mathbf{C}^4)$

$$\mathbf{v} = \begin{pmatrix} i \\ 1 \\ -1 \\ -i \end{pmatrix}.$$

Find the Euclidean norm and then normalize the vector.

Solution 1. We have

$$\mathbf{v}^*\mathbf{v} = (-i \ 1 \ -1 \ i) \begin{pmatrix} i \\ 1 \\ -1 \\ -i \end{pmatrix} = 1 + 1 + 1 + 1 = 4.$$

Thus $\|\mathbf{v}\| = 2$ and the normalized vector is

$$\frac{1}{2} \begin{pmatrix} i \\ 1 \\ -1 \\ -i \end{pmatrix}.$$

Problem 2. Consider the 4×4 matrix (Hamilton operator)

$$\hat{H} = \frac{\hbar\omega}{2}(\sigma_x \otimes \sigma_x - \sigma_y \otimes \sigma_y)$$

where ω is the frequency and $\hbar$ is the Planck constant divided by 2π. Find the *norm* of $\hat{H}$, i.e.,

$$\|\hat{H}\| := \max_{\|\mathbf{x}\|=1} \|\hat{H}\mathbf{x}\|, \qquad \mathbf{x} \in \mathbf{C}^4$$

applying two different methods. In the first method apply the *Lagrange multiplier method*, where the constraint is $\|\mathbf{x}\| = 1$. In the second method we calculate $\hat{H}^*\hat{H}$ and find the square root of the largest eigenvalue. This is then $\|\hat{H}\|$. Note that $\hat{H}^*\hat{H}$ is positive semi-definite.

Solution 2. In the first method we use the *Lagrange multiplier method*, where the constraint $\|\mathbf{x}\| = 1$ can be written as

$$x_1^2 + x_2^2 + x_3^2 + x_4^2 = 1.$$

Since

$$\sigma_x \otimes \sigma_x = \begin{pmatrix} 0 & 0 & 0 & 1 \\ 0 & 0 & 1 & 0 \\ 0 & 1 & 0 & 0 \\ 1 & 0 & 0 & 0 \end{pmatrix}, \quad \sigma_y \otimes \sigma_y = \begin{pmatrix} 0 & 0 & 0 & -1 \\ 0 & 0 & 1 & 0 \\ 0 & 1 & 0 & 0 \\ -1 & 0 & 0 & 0 \end{pmatrix}$$

we have

$$\hat{H} = \hbar\omega \begin{pmatrix} 0 & 0 & 0 & 1 \\ 0 & 0 & 0 & 0 \\ 0 & 0 & 0 & 0 \\ 1 & 0 & 0 & 0 \end{pmatrix}.$$

Let $\mathbf{x} = (x_1, x_2, x_3, x_4)^T \in \mathbf{C}^4$. We maximize

$$f(\mathbf{x}) := \|\hat{H}\mathbf{x}\|^2 - \lambda(x_1^2 + x_2^2 + x_3^2 + x_4^2 - 1)$$

where λ is the *Lagrange multiplier*. To find the extrema we solve the equations

$$\frac{\partial f}{\partial x_1} = 2\hbar^2\omega^2 x_1 - 2\lambda x_1 = 0$$

$$\frac{\partial f}{\partial x_2} = -2\lambda x_2 = 0$$

$$\frac{\partial f}{\partial x_3} = -2\lambda x_3 = 0$$

$$\frac{\partial f}{\partial x_4} = 2\hbar^2\omega^2 x_4 - 2\lambda x_4 = 0$$

together with the constraint $x_1^2 + x_2^2 + x_3^2 + x_4^2 = 1$. These equations can be written in the matrix form

$$\begin{pmatrix} \hbar^2\omega^2 - \lambda & 0 & 0 & 0 \\ 0 & -\lambda & 0 & 0 \\ 0 & 0 & -\lambda & 0 \\ 0 & 0 & 0 & \hbar^2\omega^2 - \lambda \end{pmatrix} \begin{pmatrix} x_1 \\ x_2 \\ x_3 \\ x_4 \end{pmatrix} = \begin{pmatrix} 0 \\ 0 \\ 0 \\ 0 \end{pmatrix}.$$

If $\lambda = 0$ then $x_1 = x_4 = 0$ and $\|\hat{H}\mathbf{x}\| = 0$, which is a minimum. If $\lambda \neq 0$ then $x_2 = x_3 = 0$ and $x_1^2 + x_4^2 = 1$ so that $\|\hat{H}\mathbf{x}\| = \hbar\omega$, which is the maximum. Thus we find $\|\hat{H}\| = \hbar\omega$.

In the second method we calculate $\hat{H}^*\hat{H}$ and find the square root of the largest eigenvalue. Since $H^* = H$ we find

$$\hat{H}^*\hat{H} = \hbar^2\omega^2 \begin{pmatrix} 1 & 0 & 0 & 0 \\ 0 & 0 & 0 & 0 \\ 0 & 0 & 0 & 0 \\ 0 & 0 & 0 & 1 \end{pmatrix}.$$

Thus the maximum eigenvalue is $\hbar^2\omega^2$ (twice degenerate) and $\|\hat{H}\| = \hbar\omega$.

Problem 3. Let A be an $n \times n$ matrix over $\mathbf{R}$. The *spectral norm* is

$$\|A\|_2 := \max_{\mathbf{x} \neq 0} \frac{\|A\mathbf{x}\|_2}{\|\mathbf{x}\|_2}.$$

It can be shown that $\|A\|_2$ can also be calculated as

$$\|A\|_2 = \sqrt{\text{largest eigenvalue of } A^T A}\,.$$

Note that the eigenvalues of $A^T A$ are real and nonnegative. Let

$$A = \begin{pmatrix} 2 & 5 \\ 1 & 3 \end{pmatrix}.$$

Calculate $\|A\|_2$ using this method.

Solution 3. We have

$$A^T A = \begin{pmatrix} 5 & 13 \\ 13 & 34 \end{pmatrix}.$$

The eigenvalues of $A^T A$ are 0.0257 and 38.9743. Thus

$$\|A\|_2 = \sqrt{38.9743} = 6.2429\,.$$

Problem 4. Consider the vectors

$$\mathbf{x}_1 = \begin{pmatrix} 1 \\ 1 \\ 1 \end{pmatrix}, \quad \mathbf{x}_2 = \begin{pmatrix} 1 \\ -1 \\ 1 \end{pmatrix}, \quad \mathbf{x}_3 = \begin{pmatrix} 1 \\ -1 \\ -1 \end{pmatrix}$$

in $\mathbf{R}^3$.
(i) Show that the vectors are linearly independent.
(ii) Apply the *Gram-Schmidt orthonormalization process* to these vectors.

Solution 4. (i) The equation

$$a\mathbf{x}_1 + b\mathbf{x}_2 + c\mathbf{x}_3 = 0$$

only admits the solution $a = b = c = 0$.
(ii) Since $\mathbf{x}_1^T \mathbf{x}_1 = 3$ we set $\mathbf{y}_1 = \mathbf{x}_1/\sqrt{3}$. Since $\alpha_1 := \mathbf{y}_1^T \mathbf{x}_2 = 1/\sqrt{3}$ we set

$$\mathbf{z}_2 = \mathbf{x}_2 - \alpha_1 \mathbf{y}_1 = \begin{pmatrix} 2/3 \\ -4/3 \\ 2/3 \end{pmatrix}.$$

Next we normalize the vector $\mathbf{z}_2$ and set

$$\mathbf{y}_2 = \frac{1}{\sqrt{8/3}} \mathbf{z}_2 = \begin{pmatrix} 1/\sqrt{6} \\ -\sqrt{2}/\sqrt{3} \\ 1/\sqrt{6} \end{pmatrix}.$$

Finally we find that

$$\mathbf{y}_1^T\mathbf{x}_3 = -\frac{1}{\sqrt{3}}, \qquad \mathbf{y}_2^T\mathbf{x}_3 = \frac{\sqrt{2}}{\sqrt{3}}$$

and therefore

$$\mathbf{z}_3 = \mathbf{x}_3 - \mathbf{y}_1^T\mathbf{x}_3\mathbf{y}_1 - \mathbf{y}_1^T\mathbf{x}_3\mathbf{y}_2 = \begin{pmatrix} 1 \\ 0 \\ -1 \end{pmatrix}.$$

Normalizing yields

$$\mathbf{y}_3 = \frac{1}{\sqrt{2}} \begin{pmatrix} 1/\sqrt{2} \\ 0 \\ -1/\sqrt{2} \end{pmatrix}.$$

Thus we find the orthonormal basis

$$\mathbf{y}_1 = \begin{pmatrix} 1/\sqrt{3} \\ 1/\sqrt{3} \\ 1/\sqrt{3} \end{pmatrix}, \qquad \mathbf{y}_2 = \begin{pmatrix} 1/\sqrt{6} \\ -\sqrt{2}/\sqrt{3} \\ 1/\sqrt{6} \end{pmatrix}, \qquad \mathbf{y}_3 = \begin{pmatrix} 1/\sqrt{2} \\ 0 \\ -1/\sqrt{2} \end{pmatrix}.$$

Problem 5. Let $\{\, \mathbf{v}_j \; : \; j = 1, 2, \ldots, r \,\}$ be an orthogonal set of vectors in $\mathbf{R}^n$ with $r \leq n$. Show that

$$\left\| \sum_{j=1}^{r} \mathbf{v}_j \right\|^2 = \sum_{j=1}^{r} \|\mathbf{v}_j\|^2.$$

Solution 5. We have

$$\left\| \sum_{j=1}^{r} \mathbf{v}_j \right\|^2 = \left\langle \sum_{j=1}^{r} \mathbf{v}_j, \sum_{k=1}^{r} \mathbf{v}_k \right\rangle = \sum_{j=1}^{r} \sum_{k=1}^{r} \langle \mathbf{v}_j, \mathbf{v}_k \rangle = \sum_{j=1}^{k} \|\mathbf{v}_j\|^2$$

where we used that for the scalar product we have $\langle \mathbf{v}_j, \mathbf{v}_k \rangle \equiv \mathbf{v}_j^*\mathbf{v}_k = 0$ for $j \neq k$.

Problem 6. Consider the 2×2 matrix over $\mathbf{C}$

$$A = \begin{pmatrix} a_{11} & a_{12} \\ a_{21} & a_{22} \end{pmatrix}.$$

Find the norm of A implied by the scalar product

$$\langle A, A \rangle = \sqrt{\mathrm{tr}(AA^*)}.$$

Solution 6. Since

$$A^* = \begin{pmatrix} a_{11}^* & a_{21}^* \\ a_{12}^* & a_{22}^* \end{pmatrix}$$

we obtain

$$AA^* = \begin{pmatrix} a_{11}a_{11}^* + a_{12}a_{12}^* & a_{11}a_{21}^* + a_{12}a_{22}^* \\ a_{21}a_{11}^* + a_{22}a_{12}^* & a_{21}a_{21}^* + a_{22}a_{22}^* \end{pmatrix}$$

where a_{ij}^* is the conjugate complex of a_{ij}. Thus

$$\mathrm{tr}(A^*A) = a_{11}a_{11}^* + a_{12}a_{12}^* + a_{21}a_{21}^* + a_{22}a_{22}^*.$$

Therefore the norm of A is

$$\|A\| = \sqrt{a_{11}a_{11}^* + a_{12}a_{12}^* + a_{21}a_{21}^* + a_{22}a_{22}^*}.$$

Problem 7. Let A, B be $n \times n$ matrices over $\mathbf{C}$. A scalar product is given by

$$\langle A, B \rangle = \mathrm{tr}(AB^*).$$

Let U be a unitary $n \times n$ matrix, i.e. we have $U^{-1} = U^*$.
(i) Calculate $\langle U, U \rangle$. Then find the norm implied by the scalar product.
(ii) Calculate

$$\|U\| := \max_{\|\mathbf{x}\|=1} \|U\mathbf{x}\|.$$

Solution 7. (i) We have

$$\langle U, U \rangle = \mathrm{tr}(UU^*) = \mathrm{tr}I_n = n$$

where we used that $U^{-1} = U^*$. Thus the norm is given by $\|U\| = \sqrt{n}$.
(ii) With $\|\mathbf{x}\| = 1$ and $UU^* = I_n$ we find

$$\|U\|^2 = \mathbf{x}^*U^*U\mathbf{x} = \mathbf{x}^*\mathbf{x} = 1.$$

Problem 8. (i) Let $\{\mathbf{x}_j : j = 1, 2, \ldots, n\}$ be an orthonormal basis in $\mathbf{C}^n$. Let $\{\mathbf{y}_j : j = 1, 2, \ldots, n\}$ be another orthonormal basis in $\mathbf{C}^n$. Show that

$$(U_{jk}) := (\mathbf{x}_j^*\mathbf{y}_k)$$

is a unitary matrix, where $\mathbf{x}_j^*\mathbf{y}_k$ is the scalar product of the vectors $\mathbf{x}_j$ and $\mathbf{y}_k$. This means showing that $UU^* = I_n$.
(ii) Consider the bases in $\mathbf{C}^2$

$$\mathbf{x}_1 = \frac{1}{\sqrt{2}}\begin{pmatrix} 1 \\ 1 \end{pmatrix}, \qquad \mathbf{x}_2 = \frac{1}{\sqrt{2}}\begin{pmatrix} 1 \\ -1 \end{pmatrix}$$

and

$$\mathbf{y}_1 = \frac{1}{\sqrt{2}} \begin{pmatrix} 1 \\ i \end{pmatrix}, \qquad \mathbf{y}_2 = \frac{1}{\sqrt{2}} \begin{pmatrix} 1 \\ -i \end{pmatrix}.$$

Use these bases to construct the corresponding 2×2 unitary matrix.

Solution 8.   (i) Using the *completeness relation*

$$\sum_{\ell=1}^{n} \mathbf{y}_\ell \mathbf{y}_\ell^* = I_n$$

and the property that matrix multiplication is associative we have

$$\sum_{\ell=1}^{n} (\mathbf{x}_j^* \mathbf{y}_\ell)(\mathbf{y}_\ell^* \mathbf{x}_k) = \sum_{\ell=1}^{n} \mathbf{x}_j^* (\mathbf{y}_\ell \mathbf{y}_\ell^*) \mathbf{x}_k$$

$$= \mathbf{x}_j^* (\sum_{\ell=1}^{n} \mathbf{y}_\ell \mathbf{y}_\ell^*) \mathbf{x}_k$$

$$= \mathbf{x}_j^* I_n \mathbf{x}_k$$

$$= \mathbf{x}_j^* \mathbf{x}_k$$

$$= \delta_{jk}.$$

(ii) We have

$$u_{11} = \mathbf{x}_1^* \mathbf{y}_1 = \frac{1}{2}(1+i), \qquad u_{12} = \mathbf{x}_1^* \mathbf{y}_2 = \frac{1}{2}(1-i)$$

$$u_{21} = \mathbf{x}_2^* \mathbf{y}_1 = \frac{1}{2}(1-i), \qquad u_{22} = \mathbf{x}_2^* \mathbf{y}_2 = \frac{1}{2}(1+i).$$

Thus

$$U = \frac{1}{2} \begin{pmatrix} 1+i & 1-i \\ 1-i & 1+i \end{pmatrix}$$

with

$$U^* = U^{-1} = \frac{1}{2} \begin{pmatrix} 1-i & 1+i \\ 1+i & 1-i \end{pmatrix}.$$

Problem 9.   Find the norm $\|A\| = \sqrt{\mathrm{tr}(A^*A)}$ of the skew-hermitian matrix

$$A = \begin{pmatrix} i & 2+i \\ -2+i & 3i \end{pmatrix}$$

without calculating A^*.

Solution 9.   Since $A^* = -A$ for a skew-hermitian matrix we obtain

$$\|A\| = \sqrt{-\mathrm{tr}(A^2)}.$$

Since

$$A^2 = \begin{pmatrix} -6 & * \\ * & -14 \end{pmatrix}$$

(we only calculate the diagonal elements since we take the trace) we obtain $\|A\| = \sqrt{20}$.

Problem 10. Consider the Hilbert space $\mathcal{H}$ of the 2×2 matrices over the complex numbers with the scalar product

$$\langle A, B \rangle := \operatorname{tr}(AB^*), \qquad A, B \in \mathcal{H}.$$

Show that the rescaled Pauli matrices $\mu_j = \frac{1}{\sqrt{2}}\sigma_j$, $j = 1, 2, 3$

$$\mu_1 = \frac{1}{\sqrt{2}} \begin{pmatrix} 0 & 1 \\ 1 & 0 \end{pmatrix}, \quad \mu_2 = \frac{1}{\sqrt{2}} \begin{pmatrix} 0 & -i \\ i & 0 \end{pmatrix}, \quad \mu_3 = \frac{1}{\sqrt{2}} \begin{pmatrix} 1 & 0 \\ 0 & -1 \end{pmatrix}$$

plus the rescaled 2×2 identity matrix

$$\mu_0 = \frac{1}{\sqrt{2}} \begin{pmatrix} 1 & 0 \\ 0 & 1 \end{pmatrix}$$

form an orthonormal basis in the Hilbert space $\mathcal{H}$.

Solution 10. The Hilbert space $\mathcal{H}$ is four-dimensional. Since

$$\langle \mu_j, \mu_k \rangle = \delta_{jk}$$

and $\mu_0, \mu_1, \mu_2, \mu_3$ are nonzero matrices we have an orthonormal basis.

Problem 11. Let A and B be 2×2 diagonal matrices over **R**. Assume that

$$\operatorname{tr}(AA^T) = \operatorname{tr}(BB^T) \tag{1}$$

and

$$\max_{\|\mathbf{x}\|=1} \|A\mathbf{x}\| = \max_{\|\mathbf{x}\|=1} \|B\mathbf{x}\|. \tag{2}$$

Can we conclude that $A = B$?

Solution 11. No we cannot conclude that $A = B$, for example the two matrices

$$A = \begin{pmatrix} 1 & 0 \\ 0 & 3 \end{pmatrix}, \qquad B = \begin{pmatrix} 3 & 0 \\ 0 & 1 \end{pmatrix}$$

satisfy both conditions.

Problem 12. Let A be an $n \times n$ matrix over **C**. Let $\|.\|$ be a subordinate matrix norm for which $\|I_n\| = 1$. Assume that $\|A\| < 1$.

(i) Show that $(I_n - A)$ is nonsingular.
(ii) Show that
$$\|(I_n - A)^{-1}\| \leq (1 - \|A\|)^{-1}.$$

Solution 12.   (i) Let $\lambda_1, \lambda_2, \ldots, \lambda_n$ be the eigenvalues of A. Then the eigenvalues of $I_n - A$ are $1 - \lambda_1, 1 - \lambda_2, \ldots, 1 - \lambda_n$. Since $\|A\| < 1$, we have $|\lambda_j| < 1$ for each j. Thus, none of the numbers $1 - \lambda_1, 1 - \lambda_2, \ldots, 1 - \lambda_n$ is equal to 0. This proves that $I_n - A$ is nonsingular.
(ii) Since $\|A\| < 1$, we can write the expansion

$$(I_n - A)^{-1} = I_n + A + A^2 + \cdots = \sum_{j=0}^{\infty} A^j. \tag{1}$$

Now we have $\|A^j\| \leq \|A\|^j$. Taking the norm on both sides of (1) we have

$$\|(I_n - A)^{-1}\| \leq \sum_{j=0}^{\infty} \|A\|^j = (1 - \|A\|)^{-1}$$

since $\|I_n\| = 1$. Note that the infinite series $1 + x + x^2 + \cdots$ converges to $1/(1 - x)$ if and only if $|x| < 1$.

Problem 13.   Let A be an $n \times n$ matrix. Assume that $\|A\| < 1$. Show that
$$\|(I_n - A)^{-1} - I_n\| \leq \frac{\|A\|}{1 - \|A\|}.$$

Solution 13.   For any two $n \times n$ nonsingular matrices X, Y we have the identity
$$X^{-1} - Y^{-1} \equiv X^{-1}(Y - X)Y^{-1}.$$

If $Y = I_n$ and $X = I_n - A$, then we have

$$(I_n - A)^{-1} - I_n = (I_n - A)^{-1}A.$$

Taking the norm on both sides yields

$$\|(I_n - A)^{-1} - I_n\| \leq \|I_n - A\|^{-1}\|A\|.$$

Using $\|I_n - A\|^{-1} \leq (1 - \|A\|)^{-1}$ we obtain the result.

Problem 14.   Let A be an $n \times n$ nonsingular matrix and B an $n \times n$ matrix. Assume that $\|A^{-1}B\| < 1$.
(i) Show that $A - B$ is nonsingular.

(ii) Show that

$$\frac{\|A^{-1} - (A - B)^{-1}\|}{\|A^{-1}\|} \leq \frac{\|A^{-1}B\|}{1 - \|A^{-1}B\|}.$$

Solution 14. (i) We have

$$A - B \equiv A(I_n - A^{-1}B).$$

Since $\|A^{-1}B\| < 1$ we have that $I_n - A^{-1}B$ is nonsingular. Therefore, $A - B$ which is the product of the nonsingular matrices A and $I_n - A^{-1}B$ is also nonsingular.
(ii) Consider the identity for invertible matrices X and Y

$$X^{-1} - Y^{-1} \equiv X^{-1}(Y - X)Y^{-1}.$$

Substituting $X = A$ and $Y = A - B$, we obtain

$$A^{-1} - (A - B)^{-1} = -A^{-1}B(A - B)^{-1}.$$

Taking the norm on both sides yields

$$\|A^{-1} - (A - B)^{-1}\| \leq \|A^{-1}B\| \|(A - B)^{-1}\|.$$

For nonsingular matrices X and Y we have the identities

$$Y \equiv X - (X - Y)$$
$$\equiv X(I_n - X^{-1}(X - Y))$$

and therefore

$$Y^{-1} \equiv (I_n - X^{-1}(X - Y))^{-1}X^{-1}.$$

If we substitute $Y = A - B$ and $X = A$ we have

$$(A - B)^{-1} = (I_n - A^{-1}B)^{-1}A^{-1}.$$

Taking norms yields

$$\|(A - B)^{-1}\| \leq \|A^{-1}\| \|(I_n - A^{-1}B)^{-1}\|.$$

We know that

$$\|(I_n - A^{-1}B)^{-1}\| \leq (1 - \|A^{-1}B\|)^{-1}$$

and therefore

$$\|(A - B)^{-1}\| \leq \frac{\|A^{-1}\|}{1 - \|A^{-1}B\|}.$$

It follows that

$$\|A^{-1} - (A - B)^{-1}\| \le \frac{\|A^{-1}B\| \cdot \|A^{-1}\|}{1 - \|A^{-1}B\|}$$

or

$$\frac{\|A^{-1} - (A - B)^{-1}\|}{\|A^{-1}\|} \le \frac{\|A^{-1}B\|}{1 - \|A^{-1}B\|}.$$

Problem 15. Let A be an invertible $n \times n$ matrix over $\mathbf{R}$. Consider the linear system $A\mathbf{x} = \mathbf{b}$. The *condition number* of A is defined as

$$\mathrm{Cond}(A) := \|A\| \, \|A^{-1}\|.$$

Find the condition number for the matrix

$$A = \begin{pmatrix} 1 & 0.9999 \\ 0.9999 & 1 \end{pmatrix}$$

for the infinity norm, 1-norm and 2-norm.

Solution 15. The inverse matrix of A is given by

$$A^{-1} = 10^3 \begin{pmatrix} 5.000250013 & -4.999749987 \\ -4.999749987 & 5.000250013 \end{pmatrix}.$$

Thus for the infinity norm we obtain

$$\|A\|_\infty = 1.9999, \qquad \|A^{-1}\|_\infty = 10^4.$$

Therefore
$$\mathrm{Cond}_\infty(A) = 1.9999 \cdot 10^4.$$

For the 1-norm we obtain

$$\|A\|_1 = 1.9999, \qquad \|A^{-1}\|_1 = 10^4.$$

Therefore
$$\mathrm{Cond}_1(A) = 1.9999 \cdot 10^4.$$

Thus for the 2-norm we obtain

$$\|A\|_2 = 1.9999, \qquad \|A^{-1}\|_2 = 10^4.$$

Therefore
$$\mathrm{Cond}_2(A) = 1.9999 \cdot 10^4.$$

In the given problem the condition number is the same with respect to the norms. Note that this is not always the case.

Problem 16. Let A, B be $n \times n$ matrices over $\mathbf{R}$ and $t \in \mathbf{R}$. Let $\| \ \|$ be a matrix norm. Show that

$$\|e^{tA}e^{tB} - I_n\| \leq \exp(|t|(\|A\| + \|B\|)) - 1.$$

Solution 16. We have

$$\|e^{tA}e^{tB} - I_n\| = \|(e^{tA} - I_n) + (e^{tB} - I_n) + (e^{tA} - I_n)(e^{tB} - I_n)\|$$
$$\leq \|e^{tA} - I_n\| + \|e^{tB} - I_n\| + \|e^{tA} - I_n\| \cdot \|e^{tB} - I_n\|$$
$$\leq (\exp(|t| \|A\|) - 1) + (\exp(|t| \|B\|) - 1)$$
$$+ (\exp(|t| \|A\|) - 1)(\exp(|t| \|B\|) - 1)$$
$$= \exp(|t|(\|A\| + \|B\|)) - 1.$$

Problem 17. Let $A_1, A_2, \ldots, A_p$ be $m \times m$ matrices over $\mathbf{C}$. Then we have the inequality

$$\| \exp(\sum_{j=1}^{p} A_j) - (e^{A_1/n}e^{A_2/n} \cdots e^{A_p/n})^n\| \leq \frac{2}{n} \left(\sum_{j=1}^{p} \|A_j\| \right)^2$$

$$\times \exp\left(\frac{n+2}{n} \sum_{j=1}^{p} \|A_j\| \right)$$

and

$$\lim_{n \to \infty} (e^{A_1/n}e^{A_2/n} \cdots e^{A_p/n})^n = \exp(\sum_{j=1}^{p} A_j).$$

Let $p = 2$. Find the estimate for the 2×2 matrices

$$A_1 = \begin{pmatrix} 0 & 1 \\ 1 & 0 \end{pmatrix}, \qquad A_2 = \begin{pmatrix} 1 & 0 \\ 0 & 2 \end{pmatrix}.$$

Solution 17. We have

$$\|A_1\| = 1, \qquad \|A_2\| = 2.$$

Thus for the right-hand side we find

$$\frac{2}{n}(\sum_{j=1}^{2} \|A_j\|)^2 \exp\left(\frac{n+2}{n} \sum_{j=1}^{2} \|A_j\| \right) = \frac{2}{n}(1+2)^2 \exp\left(\frac{n+2}{n}(1+2) \right)$$

$$= \frac{18}{n} \exp\left(\frac{3(n+2)}{n} \right).$$

Chapter 11

Groups and Matrices

A *group* G is a set of objects $\{a, b, c, \ldots\}$ (not necessarily countable) together with a binary operation which associates with any ordered pair of elements a, b in G a third element ab in G (closure). The binary operation (called group multiplication) is subject to the following requirements:

1) There exists an element e in G called the *identity element* such that $eg = ge = g$ for all $g \in G$.

2) For every $g \in G$ there exists an *inverse element* g^{-1} in G such that $gg^{-1} = g^{-1}g = e$.

3) *Associative law.* The identity $(ab)c = a(bc)$ is satisfied for all $a, b, c \in G$.

If $ab = ba$ for all $a, b \in G$ we call the group *commutative*.

If G has a finite number of elements it has *finite order* $n(G)$, where $n(G)$ is the number of elements. Otherwise, G has infinite order.

Groups have matrix representations with invertible $n \times n$ matrices and matrix multiplication as group multiplication. The identity element is the identity matrix. The inverse element is the inverse matrix. An important subgroup is the set of unitary matrices, where $U^* = U^{-1}$.

Let $(G_1, *)$ and $(G_2, \circ)$ be groups. A function $f : G_1 \to G_2$ with

$$f(a * b) = f(a) \circ f(b), \qquad \text{for all } a, b \in G_1$$

is called a *homomorphism*.

Problem 1. (i) Show that the set of matrices

$$E = \begin{pmatrix} 1 & 0 \\ 0 & 1 \end{pmatrix}, \quad C_3 = \begin{pmatrix} -1/2 & -\sqrt{3}/2 \\ \sqrt{3}/2 & -1/2 \end{pmatrix}, \quad C_3^{-1} = \begin{pmatrix} -1/2 & \sqrt{3}/2 \\ -\sqrt{3}/2 & -1/2 \end{pmatrix}$$

$$\sigma_1 = \begin{pmatrix} 1 & 0 \\ 0 & -1 \end{pmatrix}, \quad \sigma_2 = \begin{pmatrix} -1/2 & -\sqrt{3}/2 \\ -\sqrt{3}/2 & 1/2 \end{pmatrix}, \quad \sigma_3 = \begin{pmatrix} -1/2 & \sqrt{3}/2 \\ \sqrt{3}/2 & 1/2 \end{pmatrix}$$

form a group G under matrix multiplication, where C_3^{-1} is the inverse matrix of C_3.

(ii) Find the determinant of all these matrices. Does the set of numbers

$$\{ \det E, \ \det C_3, \ \det C_3^{-1}, \ \det \sigma_1, \ \det \sigma_2, \ \det \sigma_3 \}$$

form a group under multiplication.

(iii) Find two proper subgroups.

(iv) Find the right coset decomposition. Find the left coset decomposition. We obtain the right *coset decomposition* as follows: Let G be a finite group of order g having a proper subgroup $\mathcal{H}$ of order h. Take some element g_2 of G which does not belong to the subgroup $\mathcal{H}$, and make a right coset $\mathcal{H}g_2$. If $\mathcal{H}$ and $\mathcal{H}g_2$ do not exhaust the group G, take some element g_3 of G which is not an element of $\mathcal{H}$ and $\mathcal{H}g_2$, and make a right coset $\mathcal{H}g_3$. Continue making right cosets $\mathcal{H}g_j$ in this way. If G is a finite group, all elements of G will be exhausted in a finite number of steps and we obtain the right coset decomposition.

Solution 1. (i) Matrix multiplication reveals that the set is closed under matrix multiplication. The neutral element is the 2×2 identity matrix. Each element has exactly one inverse. The associative law holds for $n \times n$ matrices. The *group table* is given by

$\cdot$	E	C_3	C_3^{-1}	σ_1	σ_2	σ_3
E	E	C_3	C_3^{-1}	σ_1	σ_2	σ_3
C_3	C_3	C_3^{-1}	E	σ_3	σ_1	σ_2
C_3^{-1}	C_3^{-1}	E	C_3	σ_2	σ_3	σ_1
σ_1	σ_1	σ_2	σ_3	E	C_3	C_3^{-1}
σ_2	σ_2	σ_3	σ_1	C_3^{-1}	E	C_3
σ_3	σ_3	σ_1	σ_2	C_3	C_3^{-1}	E

(ii) For the determinants we find

$$\det E = \det C_3 = \det C_3^{-1} = 1$$

$$\det \sigma_1 = \det \sigma_2 = \det \sigma_3 = -1.$$

The set $\{ +1, -1 \}$ forms a commutative group under multiplication.

(iii) From the group table we see that $\{E, \sigma_1\}$ is a proper subgroup. Another proper subgroup is $\{E, C_3, C_3^{-1}\}$.

(iv) We start from the proper subgroup $\mathcal{H} = \{E, \sigma_1\}$. If we multiply the elements of $\mathcal{H}$ with σ_2 on the right we obtain the set

$$\mathcal{H}\sigma_2 = \{\sigma_2, \sigma_1\sigma_2\} = \{\sigma_2, C_3\}.$$

Similarly, we have

$$\mathcal{H}\sigma_3 = \{\sigma_3, \sigma_1\sigma_3\} = \{\sigma_3, C_3^{-1}\}.$$

We see that the six elements of the group G are just exhausted by the three right cosets so that

$$G = \mathcal{H} + \mathcal{H}\sigma_2 + \mathcal{H}\sigma_3, \qquad \mathcal{H} = \{E, \sigma_1\}.$$

The left coset decomposition is given by

$$G = \mathcal{H} + \sigma_2\mathcal{H} + \sigma_3\mathcal{H}, \qquad \mathcal{H} = \{E, \sigma_1\}$$

with

$$\sigma_2\mathcal{H} = \{\sigma_2, C_3^{-1}\}, \qquad \sigma_3\mathcal{H} = \{\sigma_3, C_3\}.$$

Problem 2. We know that the set of matrices

$$E = \begin{pmatrix} 1 & 0 \\ 0 & 1 \end{pmatrix}, \quad C_3 = \begin{pmatrix} -1/2 & -\sqrt{3}/2 \\ \sqrt{3}/2 & -1/2 \end{pmatrix}, \quad C_3^{-1} = \begin{pmatrix} -1/2 & \sqrt{3}/2 \\ -\sqrt{3}/2 & -1/2 \end{pmatrix}$$

$$\sigma_1 = \begin{pmatrix} 1 & 0 \\ 0 & -1 \end{pmatrix}, \quad \sigma_2 = \begin{pmatrix} -1/2 & -\sqrt{3}/2 \\ -\sqrt{3}/2 & 1/2 \end{pmatrix}, \quad \sigma_3 = \begin{pmatrix} -1/2 & \sqrt{3}/2 \\ \sqrt{3}/2 & 1/2 \end{pmatrix}$$

forms a group G under matrix multiplication, where C_3^{-1} is the inverse matrix of C_3. The set of matrices (3×3 *permutation matrices*)

$$I = P_0 = \begin{pmatrix} 1 & 0 & 0 \\ 0 & 1 & 0 \\ 0 & 0 & 1 \end{pmatrix}, \qquad P_1 = \begin{pmatrix} 0 & 0 & 1 \\ 1 & 0 & 0 \\ 0 & 1 & 0 \end{pmatrix},$$

$$P_2 = \begin{pmatrix} 0 & 1 & 0 \\ 0 & 0 & 1 \\ 1 & 0 & 0 \end{pmatrix} \qquad P_3 = \begin{pmatrix} 1 & 0 & 0 \\ 0 & 0 & 1 \\ 0 & 1 & 0 \end{pmatrix},$$

$$P_4 = \begin{pmatrix} 0 & 0 & 1 \\ 0 & 1 & 0 \\ 1 & 0 & 0 \end{pmatrix}, \qquad P_5 = \begin{pmatrix} 0 & 1 & 0 \\ 1 & 0 & 0 \\ 0 & 0 & 1 \end{pmatrix}$$

also forms a group G under matrix multiplication. Are the two groups isomorphic? A homomorphism which is $1-1$ and onto is an *isomorphism*.

Solution 2. Using the map

$$E \rightarrow I = P_0, \quad C_3 \rightarrow P_1, \quad C_3^{-1} \rightarrow P_2, \quad \sigma_1 \rightarrow P_3, \quad \sigma_2 \rightarrow P_4, \quad \sigma_3 \rightarrow P_5$$

we see that the two groups are isomorphic.

Problem 3. (i) Show that the matrices

$$A = \begin{pmatrix} 1 & 0 & 0 \\ 0 & 1 & 0 \\ 0 & 0 & 1 \end{pmatrix}, \quad B = \begin{pmatrix} 0 & 0 & 1 \\ 0 & 1 & 0 \\ 1 & 0 & 0 \end{pmatrix}$$

$$C = \begin{pmatrix} -1 & 0 & 0 \\ 0 & -1 & 0 \\ 0 & 0 & -1 \end{pmatrix}, \quad D = \begin{pmatrix} 0 & 0 & -1 \\ 0 & -1 & 0 \\ -1 & 0 & 0 \end{pmatrix}$$

form a group under matrix multiplication.
(ii) Show that the matrices

$$X = \begin{pmatrix} 1 & 0 & 0 & 0 \\ 0 & 1 & 0 & 0 \\ 0 & 0 & 1 & 0 \\ 0 & 0 & 0 & 1 \end{pmatrix}, \quad Y = \begin{pmatrix} 0 & 0 & 1 & 0 \\ 0 & 1 & 0 & 0 \\ 1 & 0 & 0 & 0 \\ 0 & 0 & 0 & 1 \end{pmatrix}$$

$$V = \begin{pmatrix} -1 & 0 & 0 & 0 \\ 0 & -1 & 0 & 0 \\ 0 & 0 & -1 & 0 \\ 0 & 0 & 0 & -1 \end{pmatrix}, \quad W = \begin{pmatrix} 0 & 0 & -1 & 0 \\ 0 & -1 & 0 & 0 \\ -1 & 0 & 0 & 0 \\ 0 & 0 & 0 & -1 \end{pmatrix}$$

form a group under matrix multiplication.
(iii) Show that the two groups (so-called *Vierergruppe*) are isomorphic.

Solution 3. (i) The group table is

·	A	B	C	D
A	A	B	C	D
B	B	A	D	C
C	C	D	A	B
D	D	C	B	A

The neutral element is A. Each element is its own inverse $A^{-1} = A$, $B^{-1} = B$, $C^{-1} = C$, $D^{-1} = D$.
(ii) The group table is

·	X	Y	V	W
X	X	Y	V	W
Y	Y	X	W	V
V	V	W	X	Y
W	W	V	Y	X

The neutral element is X. Each element is its own inverse $X^{-1} = X$, $Y^{-1} = Y$, $V^{-1} = V$, $W^{-1} = W$.
(iii) The map $A \leftrightarrow X$, $B \leftrightarrow Y$, $C \leftrightarrow V$, $D \leftrightarrow W$ provides the isomorphism.

Problem 4.   (i) Let $x \in \mathbf{R}$. Show that the 2×2 matrices

$$A(x) = \begin{pmatrix} 1 & x \\ 0 & 1 \end{pmatrix}$$

form a group under matrix multiplication.
(ii) Is the group commutative?
(iii) Find a group that is isomorphic to this group.

Solution 4.   (i) We have (closure)

$$A(x)A(y) = \begin{pmatrix} 1 & x \\ 0 & 1 \end{pmatrix} \begin{pmatrix} 1 & y \\ 0 & 1 \end{pmatrix} = \begin{pmatrix} 1 & x+y \\ 0 & 1 \end{pmatrix} = A(x+y).$$

The neutral element is the identity matrix I_2. The inverse element is given by

$$A(-x) = \begin{pmatrix} 1 & -x \\ 0 & 1 \end{pmatrix}.$$

Matrix multiplication is associative. Thus we have a group.
(ii) Since $A(x)A(y) = A(y)A(x)$ the group is commutative.
(iii) The group is isomorphic to the group $(\mathbf{R}, +)$.

Problem 5.   Let $a, b, c, d \in \mathbf{Z}$. Show that the 2×2 matrices

$$A = \begin{pmatrix} a & b \\ c & d \end{pmatrix}$$

with $ad - bc = 1$ form a group under matrix multiplication.

Solution 5.   We have

$$\begin{pmatrix} a_1 & b_1 \\ c_1 & d_1 \end{pmatrix} \begin{pmatrix} a_2 & b_2 \\ c_2 & d_2 \end{pmatrix} = \begin{pmatrix} a_1 a_2 + b_1 c_2 & a_1 b_2 + b_1 d_2 \\ c_1 a_2 + d_1 c_2 & c_1 b_2 + d_1 d_2 \end{pmatrix}.$$

Since $a_1, b_1, c_1, d_1 \in \mathbf{Z}$ and $a_2, b_2, c_2, d_2 \in \mathbf{Z}$ the entries of the matrix on the right-hand side are again elements in $\mathbf{Z}$. Since $\det(A_1 A_2) = \det A_1 \det A_2$ and $\det A_1 = 1$, $\det A_2 = 1$ we have $\det(A_1 A_2) = 1$. The inverse element of A is given by

$$A^{-1} = \begin{pmatrix} d & -b \\ -c & a \end{pmatrix}.$$

Thus the entries of A^{-1} are again elements in the set of integers $\mathbf{Z}$.

Problem 6. The Lie group $SU(2)$ is defined by

$$SU(2) := \{\, U \ 2 \times 2 \text{ matrix} : UU^* = I_2, \ \det U = 1 \,\}.$$

Let (3-sphere)

$$S^3 := \{\, (x_1, x_2, x_3, x_4) \in \mathbf{R}^4 \ : \ x_1^2 + x_2^2 + x_3^2 + x_4^2 = 1 \,\}.$$

Show that $SU(2)$ can be identified as a real manifold with the 3-sphere S^3.

Solution 6. Let $a, b, c, d \in \mathbf{C}$ and

$$U = \begin{pmatrix} a & b \\ c & d \end{pmatrix}.$$

Imposing the conditions $UU^* = I_2$ and $\det U = 1$ we find that U has the form

$$U = \begin{pmatrix} a & b \\ -\bar{b} & \bar{a} \end{pmatrix}$$

where $a\bar{a} + b\bar{b} = 1$. Now we embed $SU(2)$ as a subset of $\mathbf{R}^4$ by writing the complex numbers a and b in terms of their real and imaginary parts

$$a = x_1 + ix_2, \quad b = x_3 + ix_4, \qquad x_1, x_2, x_3, x_4 \in \mathbf{R}$$

whence

$$SU(2) \longrightarrow \mathbf{R}^4 \ : \ (a, b) \to (x_1, x_2, x_3, x_4).$$

The image of this map is points such that $a\bar{a} + b\bar{b} = 1$, that is

$$x_1^2 + x_2^2 + x_3^2 + x_4^2 = 1.$$

Given a and b we find

$$x_1 = \frac{a + \bar{a}}{2}, \quad x_2 = \frac{a - \bar{a}}{2i}, \quad x_3 = \frac{b + \bar{b}}{2}, \quad x_4 = \frac{b - \bar{b}}{2i}.$$

Problem 7. Let σ_1, σ_2, σ_3 be the Pauli spin matrices. Let

$$U(\alpha, \beta, \gamma) = e^{-i\alpha\sigma_3/2} e^{-i\beta\sigma_2/2} e^{-i\gamma\sigma_3/2}$$

where α, β, γ are the three *Euler angles* with the range $0 \le \alpha < 2\pi$, $0 \le \beta \le \pi$ and $0 \le \gamma < 2\pi$. Show that

$$U(\alpha, \beta, \gamma) = \begin{pmatrix} e^{-i\alpha/2} \cos(\beta/2) e^{-i\gamma/2} & -e^{-i\alpha/2} \sin(\beta/2) e^{i\gamma/2} \\ e^{i\alpha/2} \sin(\beta/2) e^{-i\gamma/2} & e^{i\alpha/2} \cos(\beta/2) e^{i\gamma/2} \end{pmatrix}. \qquad (1)$$

Solution 7. Since $\sigma_3^2 = I_2$ and $\sigma_2^2 = I_2$ we have

$$e^{-i\alpha\sigma_3/2} = I_2 \cosh(i\alpha/2) - \sigma_3 \sinh(i\alpha/2)$$
$$e^{-i\beta\sigma_2/2} = I_2 \cosh(i\beta/2) - \sigma_2 \sinh(i\beta/2)$$
$$e^{-i\gamma\sigma_3/2} = I_2 \cosh(i\gamma/2) - \sigma_3 \sinh(i\gamma/2)$$

where we used $\cosh(-z) = \cosh(z)$ and $\sinh(-z) = -\sinh(z)$. Using

$$\cosh(ix) = \cos(x), \qquad \sinh(ix) = i\sin(x)$$

we arrive at

$$U(\alpha, \beta, \gamma) = \begin{pmatrix} e^{-i\alpha/2} & 0 \\ 0 & e^{i\alpha/2} \end{pmatrix} \begin{pmatrix} \cos(\beta/2) & -\sin(\beta/2) \\ \sin(\beta/2) & \cos(\beta/2) \end{pmatrix} \begin{pmatrix} e^{-i\gamma/2} & 0 \\ 0 & e^{i\gamma/2} \end{pmatrix}.$$

Applying matrix multiplication equation (1) follows.

Problem 8. The *Heisenberg group* is the set of upper 3×3 matrices of the form

$$H = \begin{pmatrix} 1 & a & c \\ 0 & 1 & b \\ 0 & 0 & 1 \end{pmatrix}$$

where a, b, c can be taken from some (arbitrary) commutative ring.
(i) Find the inverse of H.
(ii) Given two elements x, y of a group G, we define the *commutator* of x and y, denoted by $[x, y]$ to be the element $x^{-1}y^{-1}xy$. If a, b, c are integers (in the ring $\mathbf{Z}$ of the integers) we obtain the discrete Heisenberg group H_3. It has two generators

$$x = \begin{pmatrix} 1 & 1 & 0 \\ 0 & 1 & 0 \\ 0 & 0 & 1 \end{pmatrix}, \qquad y = \begin{pmatrix} 1 & 0 & 0 \\ 0 & 1 & 1 \\ 0 & 0 & 1 \end{pmatrix}.$$

Find

$$z = xyx^{-1}y^{-1}.$$

Show that $xz = zx$ and $yz = zy$, i.e., z is the generator of the center of H_3.
(iii) The derived subgroup (or commutator subgroup) of a group G is the subgroup $[G, G]$ generated by the set of commutators of every pair of elements of G. Find $[G, G]$ for the Heisenberg group.
(iv) Let

$$A = \begin{pmatrix} 0 & a & c \\ 0 & 0 & b \\ 0 & 0 & 0 \end{pmatrix}$$

and $a, b, c \in \mathbf{R}$. Find $\exp(A)$.

(v) The Heisenberg group is a simple connected Lie group whose Lie algebra consists of matrices

$$L = \begin{pmatrix} 0 & a & c \\ 0 & 0 & b \\ 0 & 0 & 0 \end{pmatrix} .$$

Find the commutators $[L, L']$ and $[[L, L'], L']$, where $[L, L'] := LL' - L'L$.

Solution 8. (i) An inverse exists since $\det(H) = 1$. Multiplying two upper triangular matrices yields again an upper triangular matrix. From the condition

$$\begin{pmatrix} 1 & a & c \\ 0 & 1 & b \\ 0 & 0 & 1 \end{pmatrix} \begin{pmatrix} 1 & e & g \\ 0 & 1 & f \\ 0 & 0 & 1 \end{pmatrix} = \begin{pmatrix} 1 & 0 & 0 \\ 0 & 1 & 0 \\ 0 & 0 & 1 \end{pmatrix}$$

we obtain

$$e + a = 0, \quad g + af + c = 0, \quad f + b = 0 .$$

Thus

$$e = -a, \quad f = -b, \quad g = ab - c .$$

Consequently

$$H^{-1} = \begin{pmatrix} 1 & -a & ab - c \\ 0 & 1 & -b \\ 0 & 0 & 1 \end{pmatrix} .$$

(ii) Since

$$x^{-1} = \begin{pmatrix} 1 & -1 & 0 \\ 0 & 1 & 0 \\ 0 & 0 & 1 \end{pmatrix}, \quad y^{-1} = \begin{pmatrix} 1 & 0 & 0 \\ 0 & 1 & -1 \\ 0 & 0 & 1 \end{pmatrix}$$

we obtain

$$z = \begin{pmatrix} 1 & 0 & 1 \\ 0 & 1 & 0 \\ 0 & 0 & 1 \end{pmatrix}$$

and $xz = zx$, $yz = zy$. Thus z is the generator of the center of H_3.
(iii) We have

$$xyx^{-1}y^{-1} = z, \quad xzx^{-1}z^{-1} = I, \quad yzy^{-1}z^{-1} = I$$

where I is the neutral element (3×3 identity matrix) of the group. Thus the derived subgroup has only one generator z.
(iv) We use the expansion

$$e^A = \sum_{k=0}^{\infty} \frac{A^k}{k!} .$$

Since

$$A^2 = \begin{pmatrix} 0 & 0 & ab \\ 0 & 0 & 0 \\ 0 & 0 & 0 \end{pmatrix}$$

and (the matrix A is nilpotent)

$$A^n = \begin{pmatrix} 0 & 0 & 0 \\ 0 & 0 & 0 \\ 0 & 0 & 0 \end{pmatrix}$$

for $n \geq 3$ we obtain

$$e^A = \begin{pmatrix} 1 & a & c+ab/2 \\ 0 & 1 & b \\ 0 & 0 & 1 \end{pmatrix}.$$

(v) We obtain

$$[L, L'] = LL' - L'L$$

$$= \begin{pmatrix} 0 & a & c \\ 0 & 0 & b \\ 0 & 0 & 0 \end{pmatrix} \begin{pmatrix} 0 & a' & c' \\ 0 & 0 & b' \\ 0 & 0 & 0 \end{pmatrix} - \begin{pmatrix} 0 & a' & c' \\ 0 & 0 & b' \\ 0 & 0 & 0 \end{pmatrix} \begin{pmatrix} 0 & a & c \\ 0 & 0 & b \\ 0 & 0 & 0 \end{pmatrix}$$

$$= \begin{pmatrix} 0 & 0 & ab' \\ 0 & 0 & 0 \\ 0 & 0 & 0 \end{pmatrix} - \begin{pmatrix} 0 & 0 & a'b \\ 0 & 0 & 0 \\ 0 & 0 & 0 \end{pmatrix}$$

$$= \begin{pmatrix} 0 & 0 & ab' - a'b \\ 0 & 0 & 0 \\ 0 & 0 & 0 \end{pmatrix}.$$

Using this result we find the zero matrix

$$[[L, L'], L'] = \begin{pmatrix} 0 & 0 & 0 \\ 0 & 0 & 0 \\ 0 & 0 & 0 \end{pmatrix}.$$

Problem 9.   Define

$$M : \mathbf{R}^3 \to V := \{\, \mathbf{a} \cdot \sigma : \mathbf{a} \in \mathbf{R}^3 \,\} \subset \{\, 2 \times 2 \text{ complex matrices} \,\}$$
$$\mathbf{a} \to M(\mathbf{a}) = \mathbf{a} \cdot \sigma = a_1\sigma_1 + a_2\sigma_2 + a_3\sigma_3 \,.$$

This is a linear bijection between $\mathbf{R}^3$ and V. Each $U \in SU(2)$ determines a linear map $S(U)$ on $\mathbf{R}^3$ by

$$M(S(U)\mathbf{a}) = U^{-1}M(\mathbf{a})U \,.$$

The right-hand side is clearly linear in **a**. Show that $U^{-1}M(\mathbf{a})U$ is in V, that is, of the form $M(\mathbf{b})$.

Solution 9. Let

$$U = x_0 I_2 + i\mathbf{x} \cdot \boldsymbol{\sigma}$$

with $(x_0, \mathbf{x}) \in \mathbf{R}^4$ obeying $\|x_0\|^2 + \|\mathbf{x}\|^2 = 1$ and compute $U^{-1}M(\mathbf{a})U$ explicitly, where $U^{-1} = U^*$. We have

$$
\begin{aligned}
U^{-1}M(\mathbf{a})U &= (x_0 I_2 - i\mathbf{x} \cdot \boldsymbol{\sigma})(\mathbf{a} \cdot \boldsymbol{\sigma})(x_0 I_2 + i\mathbf{x} \cdot \boldsymbol{\sigma}) \\
&= (x_0 I_2 - i\mathbf{x} \cdot \boldsymbol{\sigma})(x_0 \mathbf{a} \cdot \boldsymbol{\sigma} + i(\mathbf{a} \cdot \mathbf{x})I_2 - (\mathbf{a} \times \mathbf{x}) \cdot \boldsymbol{\sigma}) \\
&= x_0^2 \mathbf{a} \cdot \boldsymbol{\sigma} + ix_0(\mathbf{a} \cdot \mathbf{x})I_2 - x_0(\mathbf{a} \times \mathbf{x}) \cdot \boldsymbol{\sigma} - ix_0(\mathbf{x} \cdot \mathbf{a})I_2 \\
&\quad + x_0(\mathbf{x} \times \mathbf{a}) \cdot \boldsymbol{\sigma} + (\mathbf{a} \cdot \mathbf{x})(\mathbf{x} \cdot \boldsymbol{\sigma}) \\
&\quad + i\mathbf{x} \cdot (\mathbf{a} \times \mathbf{x})I_2 - (\mathbf{x} \times (\mathbf{a} \times \mathbf{x})) \cdot \boldsymbol{\sigma} \\
&= x_0^2 \mathbf{a} \cdot \boldsymbol{\sigma} - 2x_0(\mathbf{a} \times \mathbf{x}) \cdot \boldsymbol{\sigma} + (\mathbf{a} \cdot \mathbf{x})(\mathbf{x} \cdot \boldsymbol{\sigma}) - (\mathbf{x} \times (\mathbf{a} \times \mathbf{x})) \cdot \boldsymbol{\sigma}
\end{aligned}
$$

since $\mathbf{x}$ is perpendicular to $\mathbf{a} \times \mathbf{x}$, where $\times$ denotes the vector product. Using the identity

$$\mathbf{c} \cdot (\mathbf{a} \times \mathbf{b}) \equiv (\mathbf{b} \cdot \mathbf{c})\mathbf{a} \quad (\mathbf{a} \cdot \mathbf{c})\mathbf{b}$$

we obtain

$$
\begin{aligned}
U^{-1}M(\mathbf{a})U &= x_0^2 \mathbf{a} \cdot \boldsymbol{\sigma} - 2x_0(\mathbf{a} \times \mathbf{x}) \cdot \boldsymbol{\sigma} + (\mathbf{a} \cdot \mathbf{x})(\mathbf{x} \cdot \boldsymbol{\sigma}) - \|\mathbf{x}\|^2 \mathbf{a} \cdot \boldsymbol{\sigma} \\
&\quad + (\mathbf{a} \cdot \mathbf{x})(\mathbf{x} \cdot \boldsymbol{\sigma}) \\
&= (x_0^2 - \|\mathbf{x}\|^2)\mathbf{a} \cdot \boldsymbol{\sigma} - 2x_0(\mathbf{a} \times \mathbf{x}) \cdot \boldsymbol{\sigma} + 2(\mathbf{a} \cdot \mathbf{x})(\mathbf{x} \cdot \boldsymbol{\sigma}).
\end{aligned}
$$

This shows, not only that $U^{-1}M(\mathbf{a})U \in V$, but also that, for

$$U = x_0 I_2 + i\mathbf{x} \cdot \boldsymbol{\sigma}$$

we have

$$S(U)\mathbf{a} = (x_0^2 - \|\mathbf{x}\|^2)\mathbf{a} + 2x_0\mathbf{x} \times \mathbf{a} + 2(\mathbf{a} \cdot \mathbf{x})\mathbf{x}.$$

Problem 10. A *topological group* G is both a group and a topological space, the two structures are related by the requirement that the maps $x \mapsto x^{-1}$ (of G onto G) and $(x, y) \mapsto xy$ (of $G \times G$ onto G) are continuous. $G \times G$ is given by the product topology.

(i) Given a topological group G, define the maps

$$\phi(x) := xax^{-1}$$

and

$$\psi(x) := xax^{-1}a^{-1} \equiv [x, a].$$

How are the iterates of the maps ϕ and ψ related?

(ii) Consider $G = SO(2)$ and

$$x = \begin{pmatrix} \cos\alpha & -\sin\alpha \\ \sin\alpha & \cos\alpha \end{pmatrix}, \qquad a = \begin{pmatrix} 0 & 1 \\ -1 & 0 \end{pmatrix}$$

with $x, a \in SO(2)$. Calculate ϕ and ψ. Discuss.

Solution 10. (i) The iterates ϕ^n and ψ^n are related by

$$\phi^n(x) = \psi^n(x)a$$

since $\phi(x) = \psi(x)a$ and by induction

$$\begin{aligned} \phi^{n+1}(x) &= (\psi^n(x)a)a(\psi^n(x)a)^{-1} \\ &= \psi^n(x)a(\psi^n(x))^{-1} \\ &= \psi^{n+1}(x)a. \end{aligned}$$

(ii) Since $\phi(x) = xax^{-1}$ we find

$$xax^{-1} = \begin{pmatrix} \cos\alpha & -\sin\alpha \\ \sin\alpha & \cos\alpha \end{pmatrix} \begin{pmatrix} 0 & 1 \\ -1 & 0 \end{pmatrix} \begin{pmatrix} \cos\alpha & \sin\alpha \\ -\sin\alpha & \cos\alpha \end{pmatrix} = \begin{pmatrix} 0 & 1 \\ -1 & 0 \end{pmatrix}.$$

Thus a is a *fixed point*, i.e. $\phi(a) = a$, of ϕ. Since $\psi(x) = xax^{-1}a^{-1}$ we obtain

$$xax^{-1}a^{-1} = \begin{pmatrix} 1 & 0 \\ 0 & 1 \end{pmatrix} = I_2.$$

Thus $\psi^n(x) = I_2$ for all n.

Problem 11. Show that the matrices

$$\begin{pmatrix} 1 & 1 \\ 0 & 1 \end{pmatrix}, \qquad \begin{pmatrix} 1 & -1 \\ 0 & 1 \end{pmatrix}$$

are conjugate in $SL(2, \mathbf{C})$ but not in $SL(2, \mathbf{R})$ (the real matrices in $SL(2, \mathbf{C})$).

Solution 11. Let

$$S = \begin{pmatrix} s_{11} & s_{12} \\ s_{21} & s_{22} \end{pmatrix}$$

with $\det S = s_{11}s_{22} - s_{12}s_{21} = 1$, i.e. $S \in SL(2, \mathbf{C})$. Now

$$\begin{aligned} S \begin{pmatrix} 1 & 1 \\ 0 & 1 \end{pmatrix} S^{-1} &= \begin{pmatrix} s_{11} & s_{12} \\ s_{21} & s_{22} \end{pmatrix} \begin{pmatrix} 1 & 1 \\ 0 & 1 \end{pmatrix} \begin{pmatrix} s_{22} & -s_{12} \\ -s_{21} & s_{11} \end{pmatrix} \\ &= \begin{pmatrix} 1 - s_{11}s_{21} & s_{11}^2 \\ -s_{21}^2 & 1 + s_{11}s_{21} \end{pmatrix} = \begin{pmatrix} 1 & -1 \\ 0 & 1 \end{pmatrix}. \end{aligned}$$

We obtain the four conditions

$$1 - s_{11}s_{21} = 1$$
$$s_{11}^2 = -1$$
$$-s_{21}^2 = 0$$
$$1 + s_{11}s_{21} = 1 .$$

It follows that $s_{21} = 0$, $s_{11} = i$. The entry s_{22} follows from $\det S = s_{11}s_{22} - s_{12}s_{21} = 1$. We have $s_{11}s_{22} = 1$ and therefore $s_{22} = -i$. The entry s_{12} is arbitrary. Thus

$$S = \begin{pmatrix} i & s_{12} \\ 0 & -i \end{pmatrix}$$

with

$$S^{-1} = \begin{pmatrix} -i & -s_{12} \\ 0 & i \end{pmatrix} .$$

Problem 12. (i) Let G be a finite set of real $n \times n$ matrices $\{ A_j \}$, $1 \leq i \leq r$, which forms a group under matrix multiplication. Suppose that

$$\text{tr}\left(\sum_{j=1}^{r} A_j\right) = \sum_{j=1}^{r} \text{tr}(A_j) = 0$$

where tr denotes the trace. Show that

$$\sum_{j=1}^{r} A_j = 0_n .$$

(ii) Show that the 2×2 matrices

$$B_1 = \begin{pmatrix} 1 & 0 \\ 0 & 1 \end{pmatrix}, \quad B_2 = \begin{pmatrix} \omega & 0 \\ 0 & \omega^2 \end{pmatrix}, \quad B_3 = \begin{pmatrix} \omega^2 & 0 \\ 0 & \omega \end{pmatrix}$$

$$B_4 = \begin{pmatrix} 0 & 1 \\ 1 & 0 \end{pmatrix}, \quad B_5 = \begin{pmatrix} 0 & \omega \\ \omega^2 & 0 \end{pmatrix}, \quad B_6 = \begin{pmatrix} 0 & \omega^2 \\ \omega & 0 \end{pmatrix}$$

form a group under matrix multiplication, where

$$\omega := \exp(2\pi i/3) .$$

(iii) Show that

$$\sum_{j=1}^{6} \text{tr}(B_j) = 0 .$$

Solution 12.   (i) Let

$$S := \sum_{i=1}^{r} A_i \,.$$

For any j, the sequence of matrices $A_j A_1, A_j A_2, \ldots, A_j A_r$ is a permutation of the elements of G, and summing yields $A_j S = S$. Thus

$$\sum_{j=1}^{r} A_j S = S \sum_{j=1}^{r} A_j = \sum_{j=1}^{r} S \Rightarrow S^2 = rS \,.$$

Therefore the minimal polynomial of S divides $x^2 - rx$, and every eigenvalue of S is either 0 or r since $x^2 - rx \equiv x(x - r)$. However the eigenvalues counted with multiplicity sum to $\mathrm{tr}(S) = 0$. Thus they are all 0. Now every eigenvalue of $S - rI_n$ is $-r \neq 0$, so the matrix $S - rI_n$ is invertible. Therefore from $S(S - rI_n) = 0_n$ we obtain $S = 0_n$.

(ii) The group table is

$\cdot$	B_1	B_2	B_3	B_4	B_5	B_6
B_1	B_1	B_2	B_3	B_4	B_5	B_6
B_2	B_2	B_3	B_1	B_5	B_6	B_4
B_3	B_3	B_1	B_2	B_6	B_4	B_5
B_4	B_4	B_6	B_5	B_1	B_3	B_2
B_5	B_5	B_4	B_6	B_2	B_1	B_3
B_6	B_6	B_5	B_4	B_3	B_2	B_1

(iii) We have

$$\sum_{j=1}^{6} \mathrm{tr} B_j = 2 + 2(\omega + \omega^2) = 0$$

where we used that $\omega + \omega^2 = -1$. The matrices B_2, B_3, B_5, B_6 contain complex numbers as entries. This is a special case of

$$\sum_{j=1}^{6} B_j = \begin{pmatrix} 0 & 0 \\ 0 & 0 \end{pmatrix} \,.$$

Problem 13.   The unitary matrices are elements of the Lie group $U(n)$. The corresponding Lie algebra $u(n)$ is the set of matrices with the condition

$$X^* = -X \,.$$

An important subgroup of $U(n)$ is the Lie group $SU(n)$ with the condition that $\det U = 1$. The unitary matrices

$$\frac{1}{\sqrt{2}} \begin{pmatrix} 1 & 1 \\ 1 & -1 \end{pmatrix}, \quad \begin{pmatrix} 0 & 1 \\ 1 & 0 \end{pmatrix}$$

are not elements of the Lie algebra $SU(2)$ since the determinants of these unitary matrices are -1. The corresponding Lie algebra $su(n)$ of the Lie group $SU(n)$ are the $n \times n$ matrices given by

$$X^* = -X, \qquad \mathrm{tr}X = 0.$$

Let σ_1, σ_2, σ_3 be the Pauli spin matrices. Then any unitary matrix in $U(2)$ can be represented by

$$U(\alpha, \beta, \gamma, \delta) = e^{i\alpha I_2} e^{-i\beta\sigma_3/2} e^{-i\gamma\sigma_2/2} e^{-i\delta\sigma_3/2}$$

where $0 \le \alpha < 2\pi$, $0 \le \beta < 2\pi$, $0 \le \gamma \le \pi$ and $0 \le \delta < 2\pi$. Calculate the right-hand side.

Solution 13. We find

$$U(\alpha, \beta, \gamma, \delta) = \begin{pmatrix} e^{i\alpha} & 0 \\ 0 & e^{i\alpha} \end{pmatrix} \begin{pmatrix} e^{-i\beta/2} & 0 \\ 0 & e^{i\beta/2} \end{pmatrix} \begin{pmatrix} \cos(\gamma/2) & -\sin(\gamma/2) \\ \sin(\gamma/2) & \cos(\gamma/2) \end{pmatrix}$$
$$\times \begin{pmatrix} e^{-i\delta/2} & 0 \\ 0 & e^{i\delta/2} \end{pmatrix}.$$

This is the *cosine-sine decomposition*. Each of the four matrices on the right-hand side is unitary and $e^{i\alpha}$ is unitary. Thus U is unitary and $\det(U) = e^{2i\alpha}$. We obtain the special case of the Lie group $SU(2)$ if $\alpha = 0$.

Problem 14. Given an orthonormal basis (column vectors) in $\mathbf{C}^N$ denoted by

$$\mathbf{x}_0, \ \mathbf{x}_1, \ldots, \ \mathbf{x}_{N-1}.$$

(i) Show that

$$U := \sum_{k=0}^{N-2} \mathbf{x}_k \mathbf{x}_{k+1}^* + \mathbf{x}_{N-1}\mathbf{x}_0^*$$

is a unitary matrix.
(ii) Find $\mathrm{tr}(U)$.
(iii) Find U^N.
(iv) Does U depend on the chosen basis? Prove or disprove.
Hint. Consider $N = 2$, the standard basis $(1,0)^T$, $(0,1)^T$ and the basis $\frac{1}{\sqrt{2}}(1,1)^T$, $\frac{1}{\sqrt{2}}(1,-1)^T$.
(v) Show that the set

$$\{ U, U^2, \ldots, U^N \}$$

forms a *commutative group* (*abelian group*) under matrix multiplication. The set is a subgroup of the group of all permutation matrices.

(vi) Assume that the set given above is the standard basis. Show that the matrix U is given by

$$U = \begin{pmatrix} 0 & 1 & 0 & \cdots & 0 \\ 0 & 0 & 1 & \cdots & 0 \\ \vdots & \vdots & \vdots & \vdots & \vdots \\ 0 & 0 & 0 & \cdots & 1 \\ 1 & 0 & 0 & \cdots & 0 \end{pmatrix}.$$

Solution 14. (i) Since $x_j^* x_k = \delta_{jk}$ we have

$$UU^* = \left(\sum_{k=0}^{N-2} x_k x_{k+1}^* + x_{N-1} x_0^* \right) \left(\sum_{k=0}^{N-2} x_{k+1} x_k^* + x_0 x_{N-1}^* \right)$$

$$= \sum_{k=0}^{N-1} x_k x_k^* = I_N.$$

(ii) Obviously we have

$$\text{tr}(U) = 0$$

since the terms $x_k x_k^*$ do not appear in the sum (i.e. we calculate the trace in the basis $x_0, \ldots, x_{N-1}$).

(iii) We notice that U maps x_k to x_{k-1}. Applying this N times and using modulo N arithmetic we obtain (i.e., U^N maps x_k to x_{k-N})

$$U^N = I_N.$$

(iv) For the standard basis in $\mathbf{C}^2$ $\{ (1,0)^T, (0,1)^T \}$ we obtain

$$U_{std} = \begin{pmatrix} 0 & 1 \\ 1 & 0 \end{pmatrix}.$$

For the basis in $\mathbf{C}^2$ $\{ \frac{1}{\sqrt{2}}(1,1)^T, \frac{1}{\sqrt{2}}(1,-1)^T \}$ we obtain

$$U_{had} = \begin{pmatrix} 1 & 0 \\ 0 & -1 \end{pmatrix}.$$

Obviously the two unitary matrices are different.

(v) Since $U_N = I_N = U^0$ we have that

$$U^s U^t = U^{s+t} = U^{s+t \bmod N}.$$

Thus the set of matrices $\{ U, U^2, \ldots, U^N \}$ forms an abelian group under matrix multiplication because $\{ 0, 1, \ldots, N-1 \}$ forms a group under addition modulo N. The two groups are isomorphic.

(vi) Let e_j denote the element of the standard basis in $\mathbf{C}^n$ with a 1 in the jth position (numbered from 0) and 0 in all other positions. Then U is given by

$$U = \sum_{k=0}^{N-2} e_k e_{k+1}^T + e_{N-1} e_0^T .$$

In the product $e_k e_{k+1}^T$, e_k denotes the row and e_{k+1}^T denotes the column in the matrix U. Thus we obtain the matrix described above.

Problem 15.   (i) Let

$$M := \frac{1}{\sqrt{2}} \begin{pmatrix} 1 & i & 0 & 0 \\ 0 & 0 & i & 1 \\ 0 & 0 & i & -1 \\ 1 & -i & 0 & 0 \end{pmatrix} .$$

Is the matrix M unitary?

(ii) Let

$$U_H := \frac{1}{\sqrt{2}} \begin{pmatrix} 1 & 1 \\ 1 & -1 \end{pmatrix}, \qquad U_S := \begin{pmatrix} 1 & 0 \\ 0 & i \end{pmatrix}$$

and

$$U_{CNOT2} = \begin{pmatrix} 1 & 0 & 0 & 0 \\ 0 & 0 & 0 & 1 \\ 0 & 0 & 1 & 0 \\ 0 & 1 & 0 & 0 \end{pmatrix} .$$

Show that the matrix M can be written as

$$M = U_{CNOT2}(I_2 \otimes U_H)(U_S \otimes U_S) .$$

(iii) Let $SO(4)$ be the special orthogonal Lie group. Let $SU(2)$ be the special unitary Lie group. Show that for every real orthogonal matrix $U \in SO(4)$, the matrix MUM^{-1} is the Kronecker product of two 2-dimensional special unitary matrices, i.e.,

$$MUM^{-1} \in SU(2) \otimes SU(2) .$$

Solution 15.   (i) Since $MM^* = I_4$ we find that M is unitary.
(ii) We obtain

$$M = \begin{pmatrix} 1 & 0 & 0 & 0 \\ 0 & 0 & 0 & 1 \\ 0 & 0 & 1 & 0 \\ 0 & 1 & 0 & 0 \end{pmatrix} \begin{pmatrix} \frac{1}{\sqrt{2}} & \frac{1}{\sqrt{2}} & 0 & 0 \\ \frac{1}{\sqrt{2}} & -\frac{1}{\sqrt{2}} & 0 & 0 \\ 0 & 0 & \frac{1}{\sqrt{2}} & \frac{1}{\sqrt{2}} \\ 0 & 0 & \frac{1}{\sqrt{2}} & -\frac{1}{\sqrt{2}} \end{pmatrix} \begin{pmatrix} 1 & 0 & 0 & 0 \\ 0 & i & 0 & 0 \\ 0 & 0 & i & 0 \\ 0 & 0 & 0 & -1 \end{pmatrix} .$$

(iii) We show that for every $A \otimes B \in SU(2) \otimes SU(2)$, we have

$$M^{-1}(A \otimes B)M \in SO(4).$$

Now every matrix $A \in SU(2)$ can be written as

$$R_z(\alpha)R_y(\theta)R_z(\beta)$$

for some $\alpha, \beta, \theta \in \mathbf{R}$, where

$$R_y(\theta) = \begin{pmatrix} \cos(\theta/2) & \sin(\theta/2) \\ -\sin(\theta/2) & \cos(\theta/2) \end{pmatrix}, \qquad R_z(\alpha) = \begin{pmatrix} e^{-i\alpha/2} & 0 \\ 0 & e^{i\alpha/2} \end{pmatrix}.$$

Therefore any matrix $A \otimes B \in SU(2) \otimes SU(2)$ can be written as a product of the matrices of the form $V \otimes I_2$ or $I_2 \otimes V$, where V is either $R_y(\theta)$ or $R_z(\alpha)$. Next we have to show that $M^{-1}(V \otimes I_2)M$ and $M^{-1}(I_2 \otimes V)M$ are in $SO(4)$. We have

$$M^{-1}(R_y(\theta) \otimes I_2)M = \begin{pmatrix} \cos(\theta/2) & 0 & 0 & -\sin(\theta/2) \\ 0 & \cos(\theta/2) & \sin(\theta/2) & 0 \\ 0 & -\sin(\theta/2) & \cos(\theta/2) & 0 \\ \sin(\theta/2) & 0 & 0 & \cos(\theta/2) \end{pmatrix}$$

$$M^{-1}(R_z(\alpha) \otimes I_2)M = \begin{pmatrix} \cos(\alpha/2) & \sin(\alpha/2) & 0 & 0 \\ -\sin(\alpha/2) & \cos(\alpha/2) & 0 & 0 \\ 0 & 0 & \cos(\alpha/2) & -\sin(\alpha/2) \\ 0 & 0 & \sin(\alpha/2) & \cos(\alpha/2) \end{pmatrix}.$$

We have similar equations for the cases of $I_2 \otimes R_y(\theta)$ and $I_2 \otimes R_z(\alpha)$. Since the mapping

$$A \otimes B \to M^{-1}(A \otimes B)M$$

is one-to-one (invertible) and the Lie groups $SU(2) \otimes SU(2)$ and $SO(4)$ have the same topological dimension, we conclude that the mapping is an isomorphism between these two Lie groups.

Problem 16.   Sometimes we parametrize the group elements of the three parameter group $SO(3)$ in terms of the *Euler angles* ψ, θ, ϕ

$$A(\psi, \theta, \phi) =$$

$$\begin{pmatrix} \cos\phi\cos\theta\cos\psi - \sin\phi\sin\psi & -\cos\phi\cos\theta\sin\psi - \sin\phi\cos\psi & \cos\phi\sin\theta \\ \sin\phi\cos\theta\cos\psi + \cos\phi\sin\psi & -\sin\phi\cos\theta\sin\psi + \cos\phi\cos\psi & \sin\phi\sin\theta \\ -\sin\theta\cos\psi & \sin\theta\sin\psi & \cos\theta \end{pmatrix}$$

with the parameters falling in the intervals

$$-\pi \leq \psi < \pi, \qquad 0 \leq \theta \leq \pi, \qquad -\pi \leq \phi < \pi.$$

Describe the shortcomings this parametrization suffers.

Solution 16. If $\theta = 0$, the matrix reduces to

$$\begin{pmatrix} \cos(\phi + \psi) & -\sin(\phi + \psi) & 0 \\ \sin(\phi + \psi) & \cos(\phi + \psi) & 0 \\ 0 & 0 & 1 \end{pmatrix}.$$

Thus only $\phi+\psi$ is determined, while if $\theta = \pi$, only $\phi-\psi$ is determined, and thus at these singular points in the parametric space, ϕ and ψ no longer define a rotation matrix uniquely. Singular points arise in any parametrization scheme for the rotation matrices. The Euler parametrization is also inappropiate by the occurrence of the singularity about the identity element (3×3 identity matrix) of the group.

Problem 17. The *octonion algebra* $\mathcal{O}$ is an 8-dimensional non-associative algebra, which is defined in terms of the basis elements e_μ ($\mu = 0, 1, \ldots, 7$) and their multiplication table. e_0 is the unit element. We use greek indices $(\mu, \nu, \ldots)$ to include the 0 and latin indices $(i, j, k, \ldots)$ when we exclude the 0. We define

$$\hat{e}_k := e_{4+k} \quad \text{for} \quad k = 1, 2, 3.$$

The multiplication rules among the basis elements of octonions e_μ can be expressed in the form

$$e_i e_j = -\delta_{ij} e_0 + \sum_{k=1}^3 \epsilon_{ijk} e_k, \quad i,j,k = 1,2,3 \tag{1}$$

and

$$-e_4 e_i = e_i e_4 = \hat{e}_i, \quad e_4 \hat{e}_i = -\hat{e}_i e_4 = e_i, \quad e_4 e_4 = -e_0$$

$$\hat{e}_i \hat{e}_j = -\delta_{ij} e_0 - \sum_{k=1}^3 \epsilon_{ijk} e_k, \quad i,j,k = 1,2,3$$

$$-\hat{e}_j e_i = e_i \hat{e}_j = -\delta_{ij} e_4 - \sum_{k=1}^3 \epsilon_{ijk} \hat{e}_k, \quad i,j,k = 1,2,3$$

where δ_{ij} is the Kronecker delta and ϵ_{ijk} is $+1$ if (ijk) is an even permutation of (123), -1 if (ijk) is an odd permutation of (123) and 0 otherwise. We can formally summarize the multiplications as

$$e_\mu e_\nu = g_{\mu\nu} e_0 + \sum_{k=1}^7 \gamma_{\mu\nu}^k e_k$$

where

$$g_{\mu\nu} = \text{diag}(1, -1, -1, -1, -1, -1, -1, -1), \qquad \gamma_{ij}^k = -\gamma_{ji}^k$$

with $\mu, \nu = 0, 1, \ldots, 7$, and $i, j, k = 1, 2, \ldots, 7$.
(i) Show that the set $\{ e_0, e_1, e_2, e_3 \}$ is a closed associative subalgebra.
(ii) Show that the octonian algebra $\mathcal{O}$ is non-associative.

Solution 17.   (i) Owing to the relation (1) we have a faithful representation using the Pauli spin matrices

$$e_0 \to \sigma_0 \equiv I_2, \quad e_j \to -i\sigma_j \quad (j = 1, 2, 3)$$

where

$$\sigma_1 = \begin{pmatrix} 0 & 1 \\ 1 & 0 \end{pmatrix}, \qquad \sigma_2 = \begin{pmatrix} 0 & -i \\ i & 0 \end{pmatrix}, \qquad \sigma_3 = \begin{pmatrix} 1 & 0 \\ 0 & -1 \end{pmatrix}.$$

We find the isomorphism

$$e_i e_j \Leftrightarrow -\sigma_i \sigma_j = -(\delta_{ij} + i \sum_{k=1}^{3} \epsilon_{ijk} \sigma_k) \Leftrightarrow -\delta_{ij} + \sum_{k=1}^{3} \epsilon_{ijk} e_k$$

(ii) The algebra is non-associative owing to the relation

$$\hat{e}_i \hat{e}_j = -\delta_{ij} e_0 - \sum_{k=1}^{3} \epsilon_{ijk} e_k, \quad i, j, k = 1, 2, 3.$$

We cannot find a matrix representation.

Problem 18.   Consider the set

$$\left\{ e = \begin{pmatrix} 1 & 0 \\ 0 & 1 \end{pmatrix}, \quad a = \begin{pmatrix} 0 & 1 \\ 1 & 0 \end{pmatrix} \right\}.$$

Then under matrix multiplication we have a group. Consider the set

$$\{ e \otimes e, \quad e \otimes a, \quad a \otimes e, \quad a \otimes a \}.$$

Does this set form a group under matrix multiplication, where $\otimes$ denotes the Kronecker product?

Solution 18.   Using the property of the Kronecker product and matrix multiplication we have

$$(e \otimes e)(e \otimes e) = e \otimes e, \qquad (e \otimes e)(e \otimes a) = e \otimes a,$$

$$(e \otimes e)(a \otimes e) = a \otimes e, \qquad (e \otimes e)(a \otimes a) = a \otimes a,$$

$$(e \otimes a)(e \otimes e) = e \otimes a, \qquad (e \otimes a)(e \otimes a) = e \otimes e,$$

$$(e \otimes a)(a \otimes e) = a \otimes a, \qquad (e \otimes a)(a \otimes a) = a \otimes e,$$

$$(a \otimes e)(e \otimes e) = a \otimes e, \qquad (a \otimes e)(e \otimes a) = a \otimes a,$$

$$(a \otimes e)(a \otimes e) = e \otimes e, \qquad (a \otimes e)(a \otimes a) = e \otimes a,$$

$$(a \otimes a)(e \otimes e) = a \otimes a, \qquad (a \otimes a)(e \otimes a) = a \otimes e,$$

$$(a \otimes a)(a \otimes e) = e \otimes a, \qquad (a \otimes a)(a \otimes a) = e \otimes e.$$

The neutral element is $(e \otimes e)$. The inverse elements are given by

$$(e \otimes e)^{-1} = e \otimes e, \qquad (e \otimes a)^{-1} = e \otimes a,$$

$$(a \otimes e)^{-1} = a \otimes e, \qquad (a \otimes a)^{-1} = a \otimes a.$$

Thus we have a group.

Problem 19. Let

$$J := \begin{pmatrix} 0 & 1 \\ -1 & 0 \end{pmatrix}.$$

(i) Find all 2×2 matrices A over $\mathbf{R}$ such that

$$A^T J A = J.$$

(ii) Do these 2×2 matrices form a group under matrix multiplication?

Solution 19. (i) From the condition

$$\begin{pmatrix} a_{11} & a_{21} \\ a_{12} & a_{22} \end{pmatrix} \begin{pmatrix} 0 & 1 \\ -1 & 0 \end{pmatrix} \begin{pmatrix} a_{11} & a_{12} \\ a_{21} & a_{22} \end{pmatrix} = \begin{pmatrix} 0 & 1 \\ -1 & 0 \end{pmatrix}$$

we obtain

$$\begin{pmatrix} 0 & a_{11}a_{22} - a_{12}a_{21} \\ -a_{11}a_{22} + a_{12}a_{21} & 0 \end{pmatrix} = \begin{pmatrix} 0 & 1 \\ -1 & 0 \end{pmatrix}.$$

Thus $\det A = 1$.

(ii) All $n \times n$ matrices A with $\det A = 1$ form a subgroup of the general linear group $GL(n, \mathbf{R})$. Note that $\det(AB) \equiv \det A \det B$. Thus if $\det A = 1$ and $\det B = 1$ we have $\det(AB) = 1$.

Problem 20. Let J be the $2n \times 2n$ matrix

$$J := \begin{pmatrix} 0_n & I_n \\ -I_n & 0_n \end{pmatrix}$$

where I_n is the $n \times n$ identity matrix and 0_n is the $n \times n$ zero matrix. Show that the $2n \times 2n$ matrices A satisfying

$$A^T J A = J$$

form a group under matrix multiplication. This group is called the *symplectic group Sp(2n)*.

Solution 20.   Let

$$A^T J A = J, \qquad B^T J B = J.$$

Then

$$(AB)^T J(AB) = B^T (A^T J A)B = B^T J B = J.$$

Obviously, $I_{2n} J I_{2n} = J$ and $\det J \neq 0$. From $A^T J A = J$ it follows that

$$\det(A^T J A) = \det(A^T) \det J \det A = \det J.$$

Thus $\det(A^T) \det A = 1$. Since $\det A^T = \det A$ we have $(\det A)^2 = 1$. From $A^T J A = J$ we obtain

$$(A^T J A)^{-1} = J^{-1}.$$

Since $J^T = J^{-1}$ we arrive at

$$A^{-1} J^{-1} (A^T)^{-1} = A^{-1} J^T (A^{-1})^T = J^T.$$

Applying the transpose we find

$$A^{-1} J (A^T)^{-1} = J.$$

Let $B = (A^{-1})^T$. Then $B^T = A^{-1}$ and $B^T J B = J$.

Problem 21.   We consider the following subgroups of the Lie group $SL(2, \mathbf{R})$

$$K := \left\{ \begin{pmatrix} \cos\theta & -\sin\theta \\ \sin\theta & \cos\theta \end{pmatrix} : \theta \in [0, 2\pi) \right\}$$

$$A := \left\{ \begin{pmatrix} r^{1/2} & 0 \\ 0 & r^{-1/2} \end{pmatrix} : r > 0 \right\}$$

$$N := \left\{ \begin{pmatrix} 1 & t \\ 0 & 1 \end{pmatrix} : t \in \mathbf{R} \right\}.$$

It can be shown that any matrix $m \in SL(2, \mathbf{R})$ can be written in a unique way as the product $m = kan$ with $k \in K$, $a \in A$ and $n \in N$. This decomposition is called *Iwasawa decomposition* and has a natural generalization to

$SL(n, \mathbf{R})$, $n \geq 3$. The notation of the subgroups comes from the fact that K is a compact subgroup, A is an abelian subgroup and N is a nilpotent subgroup of $SL(2, \mathbf{R})$. Find the Iwasawa decomposition of the matrix

$$\begin{pmatrix} \sqrt{2} & 1 \\ 1 & \sqrt{2} \end{pmatrix}.$$

Solution 21. From

$$\begin{pmatrix} \sqrt{2} & 1 \\ 1 & \sqrt{2} \end{pmatrix} = \begin{pmatrix} \cos\theta & -\sin\theta \\ \sin\theta & \cos\theta \end{pmatrix} \begin{pmatrix} r^{1/2} & 0 \\ 0 & r^{-1/2} \end{pmatrix} \begin{pmatrix} 1 & t \\ 0 & 1 \end{pmatrix}$$

we obtain

$$\begin{pmatrix} \sqrt{2} & 1 \\ 1 & \sqrt{2} \end{pmatrix} = \begin{pmatrix} r^{1/2}\cos\theta & tr^{1/2}\cos\theta - r^{-1/2}\sin\theta \\ r^{1/2}\sin\theta & tr^{1/2}\sin\theta + r^{-1/2}\cos\theta \end{pmatrix}.$$

Thus we have the four conditions

$$r^{1/2}\cos\theta = \sqrt{2}$$
$$r^{1/2}t\cos\theta - r^{-1/2}\sin\theta = 1$$
$$r^{1/2}\sin\theta = 1$$
$$r^{1/2}t\sin\theta + r^{-1/2}\cos\theta = \sqrt{2}$$

for the three unknowns r, t, θ. From the first and third conditions we obtain $\tan\theta = 1/\sqrt{2}$ and therefore

$$\sin\theta = \frac{1}{\sqrt{3}}, \qquad \cos\theta = \frac{\sqrt{2}}{\sqrt{3}}.$$

It also follows that $r^{1/2} = \sqrt{3}$. From condition (2) or (4) it finally follows that $t = 2\sqrt{2}/3$. Thus we have the decomposition

$$\begin{pmatrix} \sqrt{2} & 1 \\ 1 & \sqrt{2} \end{pmatrix} = \begin{pmatrix} \sqrt{2}/\sqrt{3} & -1/\sqrt{3} \\ 1/\sqrt{3} & \sqrt{2}/\sqrt{3} \end{pmatrix} \begin{pmatrix} \sqrt{3} & 0 \\ 0 & 1/\sqrt{3} \end{pmatrix} \begin{pmatrix} 1 & 2\sqrt{2}/3 \\ 0 & 1 \end{pmatrix}.$$

Problem 22. Let $GL(m, \mathbf{C})$ be the general linear group over $\mathbf{C}$. This Lie group consists of all nonsingular $m \times m$ matrices. Let G be a Lie subgroup of $GL(m, \mathbf{C})$. Suppose $u_1, u_2, \ldots, u_n$ is a coordinate system on G in some neighborhood of I_m, the $m \times m$ identity matrix, and that $X(u_1, u_2, \ldots, u_m)$ is a point in this neighborhood. The matrix dX of differential one-forms contains n linearly independent differential one-forms since the n-dimensional Lie group G is smoothly embedded in $GL(m, \mathbf{C})$. Consider the matrix of differential one forms

$$\Omega := X^{-1}dX, \qquad X \in G.$$

The matrix Ω of differential one forms contains n-linearly independent ones.
(i) Let A be any fixed element of G. The *left-translation* by A is given by

$$X \to AX.$$

Show that $\Omega = X^{-1}dX$ is left-invariant.
(ii) Show that

$$d\Omega + \Omega \wedge \Omega = 0$$

where $\wedge$ denotes the exterior product for matrices, i.e. we have matrix multiplication together with the *exterior product*. The exterior product is linear and satisfies

$$du_j \wedge du_k = -du_k \wedge du_j.$$

Therefore $du_j \wedge du_j = 0$ for $j = 1, 2, \ldots, n$. The exterior product is also associative.
(iii) Find dX^{-1} using $XX^{-1} = I_m$.

Solution 22.   (i) We have

$$(AX)^{-1}d(AX) = X^{-1}A^{-1}(AdX) = X^{-1}(A^{-1}A)dX = X^{-1}dX$$

since $d(AX) = AdX$.
(ii) We have

$$\Omega = X^{-1}dX$$
$$X\Omega = dX$$
$$d(X\Omega) = ddX = 0$$
$$dX \wedge \Omega + Xd\Omega = 0$$
$$(X\Omega) \wedge \Omega + Xd\Omega = 0$$
$$\Omega \wedge \Omega + d\Omega = 0.$$

(iii) We have

$$XX^{-1} = I_m$$
$$d(XX^{-1}) = 0$$
$$(dX)X^{-1} + XdX^{-1} = 0$$
$$X^{-1}(dX)X^{-1} + dX^{-1} = 0.$$

Thus

$$dX^{-1} = -X^{-1}(dX)X^{-1}.$$

Problem 23.   Consider $GL(m, \mathbf{R})$ and a Lie subgroup of it. We interpret each element X of G as a linear transformation on the vector space $\mathbf{R}^m$ of row vectors $\mathbf{v} = (v_1, v_2, \ldots, v_n)$. Thus

$$\mathbf{v} \to \mathbf{w} = \mathbf{v}X.$$

Show that $d\mathbf{w} = \mathbf{w}\Omega$.

Solution 23. We have

$$\mathbf{w} = \mathbf{v}X$$
$$d\mathbf{w} = d(\mathbf{v}X) = \mathbf{v}dX$$
$$d\mathbf{w} = (\mathbf{w}X^{-1})dX$$
$$d\mathbf{w} = \mathbf{w}(X^{-1}dX)$$
$$d\mathbf{w} = \mathbf{w}\Omega.$$

This means Ω can be interpreted as an "infinitesimal groups element".

Problem 24. Consider the Lie group $SO(2)$ consisting of the matrices

$$X = \begin{pmatrix} \cos u & -\sin u \\ \sin u & \cos u \end{pmatrix}.$$

Calculate dX and $X^{-1}dX$.

Solution 24. We have

$$dX = \begin{pmatrix} -\sin u\, du & -\cos u\, du \\ \cos u\, du & -\sin u\, du \end{pmatrix} = \begin{pmatrix} -\sin u & -\cos u \\ \cos u & -\sin u \end{pmatrix} du.$$

Since

$$X^{-1} = \begin{pmatrix} \cos u & \sin u \\ -\sin u & \cos u \end{pmatrix}$$

we obtain

$$X^{-1}dX = \begin{pmatrix} 0 & -1 \\ 1 & 0 \end{pmatrix} du$$

where we used $\cos^2 u + \sin^2 u = 1$.

Problem 25. Let n be the dimension of the Lie group G. Since the vector space of differential one-forms at the identity element is an n-dimensional vector space, there are exactly n linearly independent left invariant differential one-forms in G. Let $\sigma_1, \sigma_2, \ldots, \sigma_n$ be such a system. Consider the Lie group

$$G := \left\{ \begin{pmatrix} u_1 & u_2 \\ 0 & 1 \end{pmatrix} : u_1, u_2 \in \mathbf{R}, \ u_1 > 0 \right\}.$$

Let

$$X = \begin{pmatrix} u_1 & u_2 \\ 0 & 1 \end{pmatrix}.$$

(i) Find X^{-1} and $X^{-1}dX$. Calculate the left-invariant differential one-forms. Calculate the left-invariant volume element.

(ii) Find the right-invariant forms.

Solution 25.   (i) Since $\det X = u_1$ we obtain

$$X^{-1} = \frac{1}{u_1} \begin{pmatrix} 1 & -u_2 \\ 0 & u_1 \end{pmatrix}.$$

Moreover

$$dX = \begin{pmatrix} du_1 & du_2 \\ 0 & 0 \end{pmatrix}.$$

Thus

$$X^{-1}dX = \frac{1}{u_1} \begin{pmatrix} du_1 & du_2 \\ 0 & 0 \end{pmatrix}.$$

Hence

$$\sigma_1 = \frac{du_1}{u_1}, \qquad \sigma_2 = \frac{du_2}{u_1}$$

are left-invariant. Then the left-invariant volume element follows as

$$\sigma_1 \wedge \sigma_2 = \frac{du_1 \wedge du_2}{u_1^2}.$$

(ii) We have

$$(dX)X^{-1} = \frac{1}{u_1} \begin{pmatrix} du_1 & du_2 \\ 0 & 0 \end{pmatrix} \begin{pmatrix} 1 & -u_2 \\ 0 & u_1 \end{pmatrix}$$

$$= \frac{1}{u_1} \begin{pmatrix} du_1 & -u_2 du_1 + u_1 du_2 \\ 0 & 0 \end{pmatrix}.$$

Thus

$$\sigma_1 = \frac{du_1}{u_1}, \qquad \sigma_2 = \frac{1}{u_1}(-u_2 du_1 + u_1 du_2).$$

It follows that

$$\sigma_1 \wedge \sigma_2 = \frac{du_1 \wedge du_2}{u_1}$$

since $du_1 \wedge du_1 = 0$. The left- and right-invariant volume forms are different.

Problem 26.   Consider the Lie group consisting of the matrices

$$X = \begin{pmatrix} u_1 & u_2 \\ 0 & u_1 \end{pmatrix}, \quad u_1, u_2 \in \mathbf{R}, \quad u_1 > 0.$$

Calculate X^{-1} and $X^{-1}dX$. Find the left-invariant differential one-forms and the left-invariant volume element.

Solution 26. Since $\det X = u_1^2$ we obtain

$$X^{-1} = \frac{1}{u_1^2} \begin{pmatrix} u_1 & -u_2 \\ 0 & u_1 \end{pmatrix}.$$

Moreover

$$dX = \begin{pmatrix} du_1 & du_2 \\ 0 & du_1 \end{pmatrix}.$$

From $\Omega = X^{-1}dX$ we have

$$\Omega = \frac{1}{u_1^2} \begin{pmatrix} u_1 & -u_2 \\ 0 & u_1 \end{pmatrix} \begin{pmatrix} du_1 & du_2 \\ 0 & du_1 \end{pmatrix} = \frac{1}{u_1^2} \begin{pmatrix} u_1 du_1 & u_1 du_2 - u_2 du_1 \\ 0 & u_1 du_1 \end{pmatrix}.$$

Thus we may take

$$\sigma_1 = \frac{du_1}{u_1}, \qquad \sigma_2 = \frac{u_1 du_2 - u_2 du_1}{u_1^2}.$$

Therefore the volume form is

$$\sigma_1 \wedge \sigma = \frac{du_1 \wedge du_2}{u_1^2}$$

where we used $du_1 \wedge du_1 = 0$.

Problem 27. Let V be an $N \times N$ unitary matrix, i.e. $VV^* = I_N$. The eigenvalues of V lie on the unit circle; that is, they may be expressed in the form $\exp(i\theta_n)$, $\theta_n \in \mathbf{R}$. A function $f(V) = f(\theta_1, \ldots, \theta_N)$ is called a *class function* if f is symmetric in all its variables. Weyl gave an explicit formula for averaging class functions over the circular unitary ensemble

$$\int_{U(N)} f(V)dV =$$

$$\frac{1}{(2\pi)^N N!} \int_0^{2\pi} \cdots \int_0^{2\pi} f(\theta_1, \ldots, \theta_N) \prod_{1 \le j < k \le N} |e^{i\theta_j} - e^{i\theta_k}|^2 d\theta_1 \cdots d\theta_N.$$

Thus we integrate the function $f(V)$ over $U(N)$ by parametrizing the group by the θ_i and using *Weyl's formula* to convert the integral into an N-fold integral over the θ_i. By definition the Haar measure dV is invariant under $V \to \tilde{U}V\tilde{U}^*$, where $\tilde{U}$ is any $N \times N$ unitary matrix. The matrix V can always be diagonalized by a unitary matrix, i.e.

$$V = W \begin{pmatrix} e^{i\theta_1} & \cdots & 0 \\ \vdots & \ddots & \vdots \\ 0 & \cdots & e^{i\theta_N} \end{pmatrix} W^*$$

where W is an $N \times N$ unitary matrix. Thus the integral over V can be written as an integral over the matrix elements of W and the eigenphases θ_n. Since the measure is invariant under unitary transformations, the integral over the matrix elements of U can be evaluated straightforwardly, leaving the integral over the eigenphases. Show that for f a class function we have

$$\int_{U(N)} f(V)dV = \frac{1}{(2\pi)^N} \int_0^{2\pi} \cdots \int_0^{2\pi} f(\theta_1, \ldots, \theta_N) \det(e^{i\theta_n(n-m)})d\theta_1 \cdots d\theta_N.$$

Solution 27.   From Weyl's formula we know that

$$\int_{U(N)} f(V)dV$$

can be written as

$$\frac{1}{(2\pi)^N N!} \int_0^{2\pi} \cdots \int_0^{2\pi} f(\theta_1, \ldots, \theta_N) \prod_{1 \leq j < k \leq N} |e^{i\theta_j} - e^{i\theta_k}|^2 d\theta_1 \cdots d\theta_N.$$

We first note that

$$\prod_{1 \leq j < k \leq N} |e^{i\theta_j} - e^{i\theta_k}|^2 = \det \left(\begin{pmatrix} 1 & 1 & \cdots \\ e^{i\theta_1} & e^{i\theta_2} & \cdots \\ \vdots & \ddots & \cdots \\ e^{i(N-1)\theta_1} & e^{i(N-1)\theta_2} & \cdots \end{pmatrix} \right.$$

$$\left. \times \begin{pmatrix} 1 & e^{-i\theta_1} & e^{-2i\theta_1} & \cdots \\ 1 & e^{-i\theta_2} & e^{-2i\theta_2} & \cdots \\ \vdots & \vdots & \ddots & \cdots \\ 1 & e^{-i\theta_N} & e^{-2i\theta_N} & \cdots \end{pmatrix} \right)$$

$$= \det \left(\sum_{\ell=1}^N e^{i\theta_\ell(n-m)} \right).$$

Therefore

$$\int_{U(N)} f(\theta_1, \ldots, \theta_N)dV = \frac{1}{(2\pi)^N N!} \int_0^{2\pi} \cdots \int_0^{2\pi} f(\theta_1, \ldots, \theta_N) \times$$

$$\det \begin{pmatrix} \sum_{\ell=1}^N 1 & \sum_{\ell=1}^N e^{-i\theta_\ell} & \cdots & \sum_{\ell=1}^N e^{-(N-1)i\theta_\ell} \\ \sum_{\ell=1}^N e^{i\theta_\ell} & \sum_{\ell=1}^N 1 & \cdots & \sum_{\ell=1}^N e^{-i(N-2)\theta_\ell} \\ \vdots & \vdots & \ddots & \vdots \\ \sum_{\ell=1}^N e^{i(N-1)\theta_\ell} & \sum_{\ell=1}^N e^{i(N-2)\theta_\ell} & \cdots & \sum_{\ell=1}^N 1 \end{pmatrix} d\theta_1 \cdots d\theta_N$$

Using that f is symmetric in its arguments provides

$$\int_{U(N)} f(\theta_1,\ldots,\theta_N)dV = \frac{1}{(2\pi)^N N!}\int_0^{2\pi}\cdots\int_0^{2\pi} f(\theta_1,\ldots,\theta_N)\times$$

$$\det\begin{pmatrix} 1 & e^{-i\theta_1} & \cdots & e^{-(N-1)i\theta_1} \\ \sum_{\ell=1}^N e^{i\theta_\ell} & \sum_{\ell=1}^N 1 & \cdots & \sum_{\ell=1}^N e^{-i(N-2)\theta_\ell} \\ \vdots & \vdots & \ddots & \vdots \\ \sum_{\ell=1}^N e^{i(N-1)\theta_\ell} & \sum_{\ell=1}^N e^{i(N-2)\theta_\ell} & \cdots & \sum_{\ell=1}^N 1 \end{pmatrix}d\theta_1\cdots d\theta_N\,.$$

Subtracting $e^{i\theta_1}$ times the first row from the second row then gives

$$\int_{U(N)} f(\theta_1,\ldots,\theta_N)dV = \frac{1}{(2\pi)^N N!}\int_0^{2\pi}\cdots\int_0^{2\pi} f(\theta_1,\ldots,\theta_N)\times$$

$$\det\begin{pmatrix} 1 & e^{-i\theta_1} & \cdots & e^{-(N-1)i\theta_1} \\ \sum_{\ell=2}^N e^{i\theta_\ell} & \sum_{\ell=2}^N 1 & \cdots & \sum_{\ell=2}^N e^{-i(N-2)\theta_\ell} \\ \vdots & \vdots & \ddots & \vdots \\ \sum_{\ell=2}^N e^{i(N-1)\theta_\ell} & \sum_{\ell=2}^N e^{i(N-2)\theta_\ell} & \cdots & \sum_{\ell=2}^N 1 \end{pmatrix}d\theta_1\cdots d\theta_N\,.$$

This process is continued reducing the second row to $e^{i\theta_2}$, 1, $e^{-i\theta_2}$, $\ldots$, $e^{-(N-2)i\theta_2}$ and thus pulling out a factor of $N-1$. Then doing the same to the third row and so on. The factor of $N!$ resulting from these row manipulations cancels the $N!$ in the normalization constant of Weyl's formula.

Chapter 12

Lie Algebras and Matrices

A real Lie algebra L is a real vector space together with a bilinear map, $[,] : L \times L \to L$, called the Lie bracket, such that the following identities hold for all $a, b, c \in L$

$$[a, a] = 0$$

and the so-called *Jacobi identity*

$$[a, [b, c]] + [c, [a, b]] + [b, [c, a]] = 0.$$

It follows that $[b, a] = -[a, b]$. If $\mathcal{A}$ is an associative algebra over a field $\mathcal{F}$ (for example, the $n \times n$ matrices over $\mathbf{C}$ and matrix multiplication) with the definition

$$[a, b] := ab - ba, \quad a, b \in \mathcal{A}$$

then $\mathcal{A}$ acquires the structure of a Lie algebra.

If $X \subseteq L$ then $\langle X \rangle$ denotes the *Lie subalgebra* generated by X, that is, the smallest Lie subalgebra of L containing X. It consists of all elements obtainable from X by a finite sequence of vector space operations and Lie multiplications. A set of generators for L is a subset $X \subseteq L$ such that $L = \langle X \rangle$. If L has a finite set of generators we say that it is finitely generated.

Given two Lie algebras L_1 and L_2, a *homomorphism* of Lie algebras is a function, $f : L_1 \to L_2$, that is a linear map between vector spaces L_1 and L_2 and that preserves Lie brackets, i.e., $f([a, b]) = [f(a), f(b)]$ for all $a, b \in L_1$.

Problem 1. Consider the $n \times n$ matrices E_{ij} having 1 in the (i,j) position and 0 elsewhere, where $i,j = 1,2,\ldots,n$. Calculate the commutator. Discuss.

Solution 1. Since

$$E_{ij}E_{kl} = \delta_{jk}E_{il}$$

it follows that

$$[E_{ij}, E_{kl}] = \delta_{ik}E_{il} - \delta_{li}E_{kj} \, .$$

The coefficients are all ± 1 or 0; in particular all of them lie in the field **C**. Thus the E_{ij} are the standard basis in the vector space of all $n \times n$ matrices and thus the generators for the Lie algebra of all $n \times n$ matrices with the commutator $[A, B] := AB - BA$.

Problem 2. Show that the matrices

$$x = \begin{pmatrix} 0 & 1 \\ 0 & 0 \end{pmatrix}, \quad h = \begin{pmatrix} 1 & 0 \\ 0 & -1 \end{pmatrix}, \quad y = \begin{pmatrix} 0 & 0 \\ 1 & 0 \end{pmatrix}$$

are the generators for a Lie algebra.

Solution 2. We only have to calculate the commutators and show that the right-hand side can be represented as linear combination of the generators. We find $[x, x] = [y, y] = [z, z] = 0$ and

$$[x, h] = -2x, \quad [x, y] = h, \quad [h, y] = -2y \, .$$

Thus x, y, z are generators of a Lie algebra.

Problem 3. Consider the matrices

$$h_1 = \begin{pmatrix} 1 & 0 & 0 \\ 0 & 0 & 0 \\ 0 & 0 & 1 \end{pmatrix}, \quad h_2 = \begin{pmatrix} 0 & 0 & 0 \\ 0 & 1 & 0 \\ 0 & 0 & 0 \end{pmatrix}, \quad h_3 = \begin{pmatrix} 0 & 0 & 1 \\ 0 & 0 & 0 \\ 1 & 0 & 0 \end{pmatrix}$$

$$e = \begin{pmatrix} 0 & 1 & 0 \\ 0 & 0 & 0 \\ 0 & 1 & 0 \end{pmatrix}, \quad f = \begin{pmatrix} 0 & 0 & 0 \\ 1 & 0 & 1 \\ 0 & 0 & 0 \end{pmatrix} \, .$$

Show that the matrices form a basis of a Lie algebra.

Solution 3. We know that all $n \times n$ matrices over **R** or **C** form a Lie algebra ($gl(n, \mathbf{R})$ and $gl(n, \mathbf{C})$) under the commutator. Thus we only have to calculate the commutators and prove that they can be written as linear combinations of h_1, h_2, h_3 and e, f. We find

$$[h_1, h_2] = [h_1, h_3] = [h_2, h_3] = 0$$

$$[e, f] = h_1 + h_3 - 2h_2$$
$$[h_1, f] = [h_3, f] = -[h_2, f] = -f$$
$$[h_1, e] = [h_3, e] = -[h_2, e] = e.$$

Thus we have a basis of a Lie algebra.

Problem 4.   Let A, B be $n \times n$ matrices over $\mathbf{C}$. Calculate $\mathrm{tr}([A, B])$. Discuss.

Solution 4.   Since $\mathrm{tr}(AB) = \mathrm{tr}(BA)$ we have

$$\mathrm{tr}([A, B]) = \mathrm{tr}(AB - BA) = \mathrm{tr}(AB) - \mathrm{tr}(BA) = 0.$$

Thus the trace of the commutator of any two $n \times n$ matrices is 0.

Problem 5.   An $n \times n$ matrix X over $\mathbf{C}$ is *skew-hermitian* if

$$X^* = -X.$$

Show that the commutator of two skew-hermitian matrices is again skew-hermitian. Discuss.

Solution 5.   Let X and Y be $n \times n$ skew-hermitian matrices. Then

$$\begin{aligned}
[X, Y]^* &= (XY - YX)^* \\
&= (XY)^* - (YX)^* \\
&= Y^*X^* - X^*Y^* \\
&= YX - XY \\
&= -[X, Y].
\end{aligned}$$

Thus the skew-hermitian matrices form a Lie subalgebra of the Lie algebra $g\ell(n, \mathbf{C})$.

Problem 6.   The Lie algebra $su(m)$ consists of all $m \times m$ matrices X over $\mathbf{C}$ with the conditions

$$X^* = -X$$

(i.e. X is skew-hermitian) and $\mathrm{tr}X = 0$. Note that $\exp(X)$ is a unitary matrix. Find a basis for $su(3)$.

Solution 6.   Since we have the condition $\mathrm{tr}X = 0$ the dimension of the Lie algebra is 8. A possible basis is

$$\begin{pmatrix} 0 & i & 0 \\ i & 0 & 0 \\ 0 & 0 & 0 \end{pmatrix}, \quad \begin{pmatrix} 0 & 1 & 0 \\ -1 & 0 & 0 \\ 0 & 0 & 0 \end{pmatrix}, \quad \begin{pmatrix} i & 0 & 0 \\ 0 & -i & 0 \\ 0 & 0 & 0 \end{pmatrix}$$

$$\begin{pmatrix} 0 & 0 & i \\ 0 & 0 & 0 \\ i & 0 & 0 \end{pmatrix}, \quad \begin{pmatrix} 0 & 0 & 1 \\ i & 0 & 0 \\ -1 & 0 & 0 \end{pmatrix}, \quad \begin{pmatrix} 0 & 0 & 0 \\ 0 & 0 & i \\ 0 & i & 0 \end{pmatrix}$$

$$\begin{pmatrix} 0 & 0 & 0 \\ 0 & 0 & 1 \\ 0 & -1 & 0 \end{pmatrix}, \quad \frac{1}{\sqrt{3}} \begin{pmatrix} i & 0 & 0 \\ 0 & i & 0 \\ 0 & 0 & -2i \end{pmatrix}.$$

The eight matrices are linearly independent.

Problem 7. Any fixed element X of a Lie algebra L defines a linear transformation

$$\mathrm{ad}(X) : Z \rightarrow [X, Z] \quad \text{for any} \quad Z \in L.$$

Show that for any $K \in L$ we have

$$[\mathrm{ad}(Y), \mathrm{ad}(Z)]K = \mathrm{ad}([Y, Z])K.$$

The linear mapping ad gives a representation of the Lie algebra known as *adjoint representation*.

Solution 7. We have

$$\begin{aligned} [\mathrm{ad}(Y), \mathrm{ad}(Z)]K &= \mathrm{ad}(Y)\mathrm{ad}(Z)K - \mathrm{ad}(Z)\mathrm{ad}(Y)K \\ &= \mathrm{ad}(Y)[Z, K] - \mathrm{ad}(Z)[Y, K] \\ &= [Y, [Z, K]] - [Z, [Y, K]] \\ &= [[Y, Z], K] \\ &= \mathrm{ad}([Y, Z])K \end{aligned}$$

where we used the *Jacobi identity*

$$[Y, [Z, K]] + [K, [Y, Z]] + [Z, [K, Y]] = 0.$$

Problem 8. There is only one non-commutative Lie algebra L of dimension 2. If x, y are the generators (basis in L), then

$$[x, y] = x.$$

(i) Find the *adjoint representation* of this Lie algebra. Let v, w be two elements of a Lie algebra. Then we define

$$\mathrm{ad}v(w) := [v, w]$$

and $w\mathrm{ad}v := [v, w]$.

(ii) The *Killing form* is defined by

$$\kappa(x,y) := \mathrm{tr}(\mathrm{ad}x\,\mathrm{ad}y)$$

for all $x, y \in L$. Find the Killing form.

Solution 8.   (i) Since

$$\mathrm{ad}x(x) = [x,x] = 0, \quad \mathrm{ad}x(y) = [x,y] = x$$

and

$$\mathrm{ad}y(x) = [y,x] = -x, \quad \mathrm{ad}y(y) = [y,y] = 0$$

we obtain

$$(\,x \quad y\,)\begin{pmatrix} \mathrm{ad}x_{11} & \mathrm{ad}x_{12} \\ \mathrm{ad}x_{21} & \mathrm{ad}x_{22} \end{pmatrix} = (\,0 \quad x\,)$$

and

$$(\,x \quad y\,)\begin{pmatrix} \mathrm{ad}y_{11} & \mathrm{ad}y_{12} \\ \mathrm{ad}y_{21} & \mathrm{ad}y_{22} \end{pmatrix} = (\,-x \quad 0\,).$$

Since x, y are basis elements in L we obtain

$$\mathrm{ad}x = \begin{pmatrix} 0 & 1 \\ 0 & 0 \end{pmatrix} \qquad \mathrm{ad}y = \begin{pmatrix} -1 & 0 \\ 0 & 0 \end{pmatrix}.$$

(ii) We find

$$\kappa(x,x) = \mathrm{tr}(\mathrm{ad}x\,\mathrm{ad}x) = 0$$
$$\kappa(x,y) = \mathrm{tr}(\mathrm{ad}x\,\mathrm{ad}y) = 0$$
$$\kappa(y,x) = \mathrm{tr}(\mathrm{ad}y\,\mathrm{ad}x) = 0$$
$$\kappa(y,y) = \mathrm{tr}(\mathrm{ad}y\,\mathrm{ad}y) = 1.$$

This can be written in matrix form

$$\begin{pmatrix} 0 & 0 \\ 0 & 1 \end{pmatrix}.$$

Problem 9.   Consider the Lie algebra $L = s\ell(2, \mathcal{F})$ with $\mathrm{char}\mathcal{F} \neq 2$. Take as the standard basis for L the three matrices

$$x = \begin{pmatrix} 0 & 1 \\ 0 & 0 \end{pmatrix}, \quad y = \begin{pmatrix} 0 & 0 \\ 1 & 0 \end{pmatrix}, \quad h = \begin{pmatrix} 1 & 0 \\ 0 & -1 \end{pmatrix}.$$

(i) Find the multiplication table, i.e. the commutators.
(ii) Find the adjoint representation of L with the ordered basis $\{\,x\ h\ y\,\}$.

(iii) Show that L is *simple*. If L has no ideals except itself and 0, and if moreover $[L, L] \neq 0$, we call L simple. A subspace I of a Lie algebra L is called an *ideal* of L if $x \in L$, $y \in I$ together imply $[x, y] \in I$.

Solution 9. (i) We have

$$[x, y] = h, \quad [h, x] = 2x, \quad [h, y] = -2y.$$

(ii) We find

$$\mathrm{ad}(x) = \begin{pmatrix} 0 & -2 & 0 \\ 0 & 0 & 1 \\ 0 & 0 & 0 \end{pmatrix}$$

$$\mathrm{ad}(h) = \begin{pmatrix} 2 & 0 & 0 \\ 0 & 0 & 0 \\ 0 & 0 & -2 \end{pmatrix}$$

$$\mathrm{ad}(y) = \begin{pmatrix} 0 & 0 & 0 \\ -1 & 0 & 0 \\ 0 & 2 & 0 \end{pmatrix}.$$

Notice that x, y, h are eigenvectors for $\mathrm{ad}h$, corresponding to the eigenvalues $2, -2, 0$. Since $\mathrm{char}\mathcal{F} \neq 2$, these eigenvalues are distinct.

(iii) If $I \neq 0$ is an ideal of L, let $ax + by + ch$ be an arbitrary nonzero element of I. Applying $\mathrm{ad}x$ twice, we obtain $-2bx \in I$, and applying $\mathrm{ad}y$ twice, we obtain $-2ay \in I$. Therefore, if a or b is nonzero, the subset I contains either y or x ($\mathrm{char}\mathcal{F} \neq 2$), and then, clearly $I = L$ follows. On the other hand, if $a = b = 0$, then $0 \neq ch \in I$, so $h \in I$, and again $I = L$ follows. Thus we conclude that L is simple.

Problem 10. Consider the Lie algebra $gl_2(\mathbf{R})$. The matrices

$$e_1 = \begin{pmatrix} 1 & 0 \\ 0 & 0 \end{pmatrix}, \quad e_2 = \begin{pmatrix} 0 & 1 \\ 0 & 0 \end{pmatrix}, \quad e_3 = \begin{pmatrix} 0 & 0 \\ 1 & 0 \end{pmatrix}, \quad e_4 = \begin{pmatrix} 0 & 0 \\ 0 & 1 \end{pmatrix}$$

form a basis of $gl_2(\mathbf{R})$. Find the adjoint representation.

Solution 10. Since

$$\mathrm{ad}e_1(e_1) = [e_1, e_1] = 0, \qquad \mathrm{ad}e_1(e_2) = [e_1, e_2] = e_2,$$

$$\mathrm{ad}e_1(e_3) = [e_1, e_3] = -e_3, \qquad \mathrm{ad}e_1(e_4) = [e_1, e_4] = 0$$

we have

$$(e_1 \ e_2 \ e_3 \ e_4)\mathrm{ad}e_1 = (0 \ e_2 \ -e_3 \ 0).$$

Therefore

$$ad(e_1) = \begin{pmatrix} 0 & 0 & 0 & 0 \\ 0 & 1 & 0 & 0 \\ 0 & 0 & -1 & 0 \\ 0 & 0 & 0 & 0 \end{pmatrix}.$$

Analogously, we have

$$ad(e_2) = \begin{pmatrix} 0 & 0 & 1 & 0 \\ -1 & 0 & 0 & 1 \\ 0 & 0 & 0 & 0 \\ 0 & 0 & -1 & 0 \end{pmatrix}$$

$$ad(e_3) = \begin{pmatrix} 0 & -1 & 0 & 0 \\ 0 & 0 & 0 & 0 \\ 1 & 0 & 0 & -1 \\ 0 & 1 & 0 & 0 \end{pmatrix}$$

$$ad(e_4) = \begin{pmatrix} 0 & 0 & 0 & 0 \\ 0 & -1 & 0 & 0 \\ 0 & 0 & 1 & 0 \\ 0 & 0 & 0 & 0 \end{pmatrix}.$$

Problem 11.   Let $\{\, e, f \,\}$ with

$$e = \begin{pmatrix} 0 & 1 \\ 0 & 0 \end{pmatrix}, \qquad f = \begin{pmatrix} -1 & 0 \\ 0 & 0 \end{pmatrix}$$

a basis for a Lie algebra. We have $[e, f] = e$. Is $\{\, e \otimes I_2, f \otimes I_2 \,\}$ a basis of a Lie algebra? Here I_2 denotes the 2×2 unit matrix and $\otimes$ the Kronecker product.

Solution 11.   Let A, B, C, D be $n \times n$ matrices. Then we have

$$[A \otimes B, C \otimes D] \equiv (A \otimes B)(C \otimes D) - (C \otimes D)(A \otimes B)$$
$$\equiv (AC) \otimes (BD) - (CA) \otimes (DB).$$

Using this identity we obtain

$$[e \otimes I_2, f \otimes I_2] = (ef) \otimes I_2 - (fe) \otimes I_2 = (ef - fe) \otimes I_2 = e \otimes I_2$$
$$[e \otimes I_2, e \otimes I_2] = (ee) \otimes I_2 - (ee) \otimes I_2 = (ee - ee) \otimes I_2 = 0_2 \otimes 0_2$$
$$[f \otimes I_2, f \otimes I_2] = (ff) \otimes I_2 - (ff) \otimes I_2 = (ff - ff) \otimes I_2 = 0_2 \otimes 0_2.$$

Thus $\{\, e \otimes I_2, f \otimes I_2 \,\}$ is a basis of a Lie algebra.

Problem 12.   Let $\{\, e, f \,\}$ with

$$e = \begin{pmatrix} 0 & 1 \\ 0 & 0 \end{pmatrix}, \qquad f = \begin{pmatrix} -1 & 0 \\ 0 & 0 \end{pmatrix}$$

a basis for a Lie algebra. We have $[e, f] = e$. Is $\{\, e \otimes e, e \otimes f, f \otimes e, f \otimes f \,\}$ a basis of a Lie algebra?

Solution 12. Straightforward calculation yields

$$[e \otimes e, e \otimes e] = 0_4$$
$$[e \otimes e, e \otimes f] = 0_4$$
$$[e \otimes e, f \otimes e] = 0_4$$
$$[e \otimes f, e \otimes f] = 0_4$$
$$[e \otimes f, f \otimes e] = 0_4$$
$$[e \otimes f, f \otimes f] = -e \otimes f$$
$$[f \otimes e, f \otimes e] = 0_4$$
$$[f \otimes e, f \otimes f] = -f \otimes e$$
$$[f \otimes f, f \otimes f] = 0_4$$

where 0_4 is the 4×4 zero matrix. Thus the set is a basis of a Lie algebra.

Problem 13. The elements (generators) $Z_1, Z_2, \ldots, Z_r$ of an r-dimensional Lie algebra satisfy the conditions

$$[Z_\mu, Z_\nu] = \sum_{\tau=1}^{r} c_{\mu\nu}^{\tau} Z_\tau$$

with $c_{\mu\nu}^{\tau} = -c_{\nu\mu}^{\tau}$, where the $c_{\mu\nu}^{\tau}$'s are called the *structure constants*. Let A be an arbitrary linear combination of the elements

$$A = \sum_{\mu=1}^{r} a^\mu Z_\mu \,.$$

Suppose that X is some other linear combination such that

$$X = \sum_{\nu=1}^{r} b^\nu Z_\nu$$

and

$$[A, X] = \rho X \,.$$

This equation has the form of an eigenvalue equation, where ρ is the corresponding eigenvalue and X the corresponding eigenvector. Assume that the Lie algebra is represented by matrices. Find the secular equation for the eigenvalues ρ.

Solution 13. From $[A, X] = \rho X$ we have $AX - XA = \rho X$. Thus

$$\sum_{\mu=1}^{r} \sum_{\nu=1}^{r} (a^\mu b^\nu Z_\mu Z_\nu - a^\mu b^\nu Z_\nu Z_\mu) = \rho \sum_{\nu=1}^{r} b^\nu Z_\nu \,. \tag{1}$$

Inserting

$$Z_\mu Z_\nu = \sum_{\tau=1}^{r} c_{\mu\nu}^\tau Z_\tau + Z_\nu Z_\mu$$

into (1) yields

$$\sum_{\tau=1}^{r} \sum_{\nu=1}^{r} \sum_{\mu=1}^{r} a^\mu b^\nu c_{\mu\nu}^\tau Z_\tau = \rho \sum_{\tau=1}^{r} b^\tau Z_\tau .$$

It follows that

$$\sum_{\tau=1}^{r} \left(\sum_{\nu=1}^{r} \sum_{\mu=1}^{r} a^\mu b^\nu c_{\mu\nu}^\tau - \rho b^\tau \right) Z_\tau = 0 .$$

Since Z_τ, with $\tau = 1, 2, \ldots, r$ are linearly independent we have for a fixed $\tau = \tau^*$

$$\sum_{\nu=1}^{r} \left(\sum_{\mu=1}^{r} a^\mu b^\nu c_{\mu\nu}^{\tau^*} \right) - \rho b^{\tau^*} = 0 .$$

Introducing the Kronecker delta $\delta_\nu^{\tau^*}$ with 1 if $\nu = \tau^*$ and 0 otherwise we can write

$$\sum_{\nu=1}^{r} \left(\sum_{\mu=1}^{r} a^\mu c_{\mu\nu}^{\tau^*} - \rho \delta_\nu^{\tau^*} \right) b^\nu = 0 .$$

For $\nu, \tau^* = 1, 2, \ldots, r$

$$\sum_{\mu=1}^{r} a^\mu c_{\mu\nu}^{\tau^*} - \rho \delta_\nu^{\tau^*}$$

defines an $r \times r$ matrix with the secular equation

$$\det \left(\sum_{\mu=1}^{r} a^\mu c_{\mu\nu}^{\tau^*} - \rho \delta_\nu^{\tau^*} \right) = 0$$

to find the eigenvalues ρ, where $\nu, \tau^* = 1, 2, \ldots, r$.

Problem 14.   Let $c_{\sigma\lambda}^\tau$ be the structure constants of a Lie algebra. We define

$$g_{\sigma\lambda} = g_{\lambda\sigma} = \sum_{\rho=1}^{r} \sum_{\tau=1}^{r} c_{\sigma\rho}^\tau c_{\lambda\tau}^\rho$$

and

$$g^{\sigma\lambda} g_{\sigma\lambda} = \delta_\sigma^\lambda .$$

A Lie algebra L is called *semisimple* if and only if $\det |g_{\sigma\lambda}| \neq 0$. We assume in the following that the Lie algebra is semisimple. We define

$$C := \sum_{\rho=1}^{r} \sum_{\sigma=1}^{r} g^{\rho\sigma} X_\rho X_\sigma .$$

The operator C is called *Casimir operator*. Let X_τ be an element of the Lie algebra L. Calculate the commutator $[C, X_\tau]$.

Solution 14. We have

$$[C, X_\tau] = \sum_{\rho=1}^{r} \sum_{\sigma=1}^{r} g^{\rho\sigma} [X_\rho X_\sigma, X_\tau]$$

$$= \sum_{\rho=1}^{r} \sum_{\sigma=1}^{r} g^{\rho\sigma} X_\rho [X_\sigma, X_\tau] + \sum_{\rho=1}^{r} \sum_{\sigma=1}^{r} \sum_{\lambda=1}^{r} g^{\rho\sigma} c_{\rho\tau}^{\lambda} X_\lambda X_\sigma$$

$$= \sum_{\rho=1}^{r} \sum_{\sigma=1}^{r} \sum_{\lambda=1}^{r} g^{\rho\sigma} c_{\sigma\tau}^{\lambda} X_\rho X_\lambda + \sum_{\rho=1}^{r} \sum_{\sigma=1}^{r} \sum_{\lambda=1}^{r} g^{\rho\sigma} c_{\rho\tau}^{\lambda} X_\lambda X_\sigma$$

$$= \sum_{\rho=1}^{r} \sum_{\sigma=1}^{r} \sum_{\lambda=1}^{r} g^{\rho\sigma} c_{\sigma\tau}^{\lambda} X_\rho X_\lambda + \sum_{\rho=1}^{r} \sum_{\sigma=1}^{r} \sum_{\lambda=1}^{r} g^{\sigma\rho} c_{\sigma\tau}^{\lambda} X_\lambda X_\rho$$

$$= \sum_{\rho=1}^{r} \sum_{\sigma=1}^{r} \sum_{\lambda=1}^{r} g^{\rho\sigma} c_{\sigma\tau}^{\lambda} (X_\rho X_\lambda + X_\lambda X_\rho)$$

where we made a change in the variables σ and ρ. For semisimple Lie algebra we have

$$c_{\sigma\tau}^{\lambda} = \sum_{\nu=1}^{r} g^{\lambda\nu} c_{\nu\sigma\tau}$$

and therefore we obtain

$$[C, X_\tau] = \sum_{\rho=1}^{r} \sum_{\sigma=1}^{r} \sum_{\lambda=1}^{r} \sum_{\nu=1}^{r} g^{\rho\sigma} g^{\lambda\nu} c_{\nu\sigma\tau} (X_\rho X_\lambda + X_\lambda X_\rho) \,.$$

Now $c_{\nu\sigma\tau}$ is antisymmetric, while the quantity in parentheses is symmetric in ρ and λ, and hence the right-hand side must vanish to give

$$[C, X_\tau] = 0 \qquad \text{for all} \qquad X_\tau \in L \,.$$

Problem 15. Show that the matrices

$$J_x = \begin{pmatrix} 0 & 0 & 0 \\ 0 & 0 & -1 \\ 0 & 1 & 0 \end{pmatrix}, \quad J_y = \begin{pmatrix} 0 & 0 & 1 \\ 0 & 0 & 0 \\ -1 & 0 & 0 \end{pmatrix}, \quad J_z = \begin{pmatrix} 0 & -1 & 0 \\ 1 & 0 & 0 \\ 0 & 0 & 0 \end{pmatrix}$$

form generators of a Lie algebra. Is the Lie algebra simple? A Lie algebra is *simple* if it contains no ideals other than L and 0.

Solution 15. We find the commutators

$$[J_x, J_y] = J_z, \qquad [J_y, J_z] = J_x, \qquad [J_z, J_x] = J_y \,.$$

The Lie algebra contains no ideals other than L and 0. Thus the given Lie algebra is simple.

Problem 16.   The *roots* of a semisimple Lie algebra are the Lie algebra weights occurring in its adjoint representation. The set of roots forms the root system, and is completely determined by the semisimple Lie algebra. Consider the semisimple Lie algebra $s\ell(2, \mathbf{R})$ with the generators

$$H = \begin{pmatrix} 1 & 0 \\ 0 & -1 \end{pmatrix}, \qquad X = \begin{pmatrix} 0 & 1 \\ 0 & 0 \end{pmatrix}, \qquad Y = \begin{pmatrix} 0 & 0 \\ 1 & 0 \end{pmatrix}.$$

Find the roots.

Solution 16.   We obtain

$$\mathrm{ad}H(X) = [H, X] = 2X$$
$$\mathrm{ad}H(Y) = [H, Y] = -2Y$$

and $[X, Y] = H$. Thus there are two roots of $s\ell(2, \mathbf{R})$ given by $\alpha(H) = 2$ and $\alpha(H) = -2$. The Lie algebraic rank of $s\ell(2, \mathbf{R})$ is one, and it has one positive root.

Problem 17.   The Lie algbra $sl(2, \mathbf{R})$ is spanned by the matrices

$$h = \begin{pmatrix} 1 & 0 \\ 0 & -1 \end{pmatrix}, \qquad e = \begin{pmatrix} 0 & 1 \\ 0 & 0 \end{pmatrix}, \qquad f = \begin{pmatrix} 0 & 0 \\ 1 & 0 \end{pmatrix}.$$

(i) Find the commutators $[h, e]$, $[h, f]$ and $[e, f]$.
(ii) Consider

$$C = \frac{1}{2}h^2 + ef + fe.$$

Find C. Calculate the commutators $[C, h]$, $[C, e]$, $[C, f]$. Show that C can be written in the form

$$C = \frac{1}{2}h^2 + h + 2fe.$$

(iii) Consider the vector

$$\mathbf{v} = \begin{pmatrix} 1 \\ 0 \end{pmatrix}.$$

Calculate $h\mathbf{v}$, $e\mathbf{v}$, $f\mathbf{v}$ and $C\mathbf{v}$. Give an interpretation.

Solution 17.   (i) We obtain

$$[h, e] = 2e, \qquad [h, f] = -2f, \qquad [e, f] = h.$$

(ii) Straightforward calculation yields

$$C = \frac{1}{2}h^2 + ef + fe$$

$$= \frac{1}{2}\begin{pmatrix} 1 & 0 \\ 0 & 1 \end{pmatrix} + \begin{pmatrix} 0 & 1 \\ 0 & 0 \end{pmatrix}\begin{pmatrix} 0 & 0 \\ 1 & 0 \end{pmatrix} + \begin{pmatrix} 0 & 0 \\ 1 & 0 \end{pmatrix}\begin{pmatrix} 0 & 1 \\ 0 & 0 \end{pmatrix}$$

$$= \frac{1}{2}\begin{pmatrix} 1 & 0 \\ 0 & 1 \end{pmatrix} + \begin{pmatrix} 1 & 0 \\ 0 & 0 \end{pmatrix} + \begin{pmatrix} 0 & 0 \\ 0 & 1 \end{pmatrix}$$

$$= \frac{3}{2}\begin{pmatrix} 1 & 0 \\ 0 & 1 \end{pmatrix}.$$

Since C is the identity matrix times $3/2$ we find

$$[C, e] = 0_n, \qquad [C, f] = 0_n, \qquad [C, h] = 0_n.$$

Since

$$ef = fe + [e, f] = fe + h$$

we obtain

$$C = \frac{1}{2}h^2 + h + 2fe.$$

(iii) We find

$$h\mathbf{v} = \mathbf{v}, \quad e\mathbf{v} = \begin{pmatrix} 0 \\ 0 \end{pmatrix}, \quad f\mathbf{v} = \begin{pmatrix} 0 \\ 1 \end{pmatrix}, \quad C\mathbf{v} = \frac{3}{2}\mathbf{v}.$$

Thus $h\mathbf{v} = \mathbf{v}$ is an eigenvalue equation with eigenvalue $+1$. $e\mathbf{v} = \mathbf{0}$ is also an eigenvalue equation with eigenvalue 0 and finally $C\mathbf{v} = \frac{3}{2}\mathbf{v}$ is an eigenvalue equation with eigenvalue $3/2$.

Problem 18. Let L be a finite dimensional Lie algebra. Let $C^\infty(S^1)$ be the set of all infinitely differentiable functions, where S^1 is the unit circle manifold. In the product space $L \otimes C^\infty(S^1)$ we define the Lie bracket ($g_1, g_2 \in L$ and $f_1, f_2 \in C^\infty(S^1)$)

$$[g_1 \otimes f_1, g_2 \otimes f_2] := [g_1, g_2] \otimes (f_1 f_2).$$

Calculate

$$[g_1 \otimes f_1, [g_2 \otimes f_2, g_3 \otimes f_3]] + [g_3 \otimes f_3, [g_1 \otimes f_1, g_2 \otimes f_2]] + [g_2 \otimes f_2, [g_3 \otimes f_3, g_1 \otimes f_1]].$$

Solution 18. We have

$$[g_1 \otimes f_1, [g_2 \otimes f_2, g_3 \otimes f_3]] = [g_1, [g_2, g_3]] \otimes f_1 f_2 f_3$$
$$[g_3 \otimes f_3, [g_1 \otimes f_1, g_2 \otimes f_2]] = [g_3, [g_1, g_2]] \otimes f_3 f_2 f_1$$

$$[g_2 \otimes f_2, [g_3 \otimes f_3, g_1 \otimes f_1]] = [g_2, [g_3, g_1]] \otimes f_2 f_3 f_1 .$$

Since $f_1 f_2 f_3 = f_3 f_1 f_2 = f_2 f_3 f_1$ we obtain

$$([g_1, [g_2, g_3]] + [g_3, [g_1, g_2]] + [g_2, [g_3, g_1]]) \otimes f_1 f_2 f_3 = 0$$

where we applied the *Jacobi identity*.

Problem 19. A basis for the Lie algebra $su(N)$, for odd N, may be built from two unitary unimodular $N \times N$ matrices

$$g = \begin{pmatrix} 1 & 0 & 0 & \cdots & 0 \\ 0 & \omega & 0 & \cdots & 0 \\ 0 & 0 & \omega^2 & \cdots & 0 \\ \vdots & \vdots & \vdots & \ddots & \vdots \\ 0 & 0 & 0 & \cdots & \omega^{N-1} \end{pmatrix}, \qquad h = \begin{pmatrix} 0 & 1 & 0 & \cdots & 0 \\ 0 & 0 & 1 & \cdots & 0 \\ \vdots & \vdots & \vdots & \ddots & \vdots \\ 0 & 0 & 0 & \cdots & 1 \\ 1 & 0 & 0 & \cdots & 0 \end{pmatrix}$$

where ω is a primitive Nth root of unity, i.e. with period not smaller than N, here taken to be $\exp(4\pi i/N)$. We obviously have

$$hg = \omega gh . \tag{1}$$

(i) Find g^N and h^N.
(ii) Find tr g.
(iii) Let $\mathbf{m} = (m_1, m_2)$, $\mathbf{n} = (n_1, n_2)$ and define

$$\mathbf{m} \times \mathbf{n} := m_1 n_2 - m_2 n_1 .$$

The complete set of unitary unimodular $N \times N$ matrices

$$J_{m_1, m_2} := e^{m_1 m_2/2} g^{m_1} h^{m_2}$$

suffice to span the Lie algebra $su(N)$. Find J^*.
(iv) Calculate $J_\mathbf{m} J_\mathbf{n}$.
(v) Find the commutator $[J_\mathbf{m}, J_\mathbf{n}]$.

Solution 19. (i) Since $\omega^N = 1$ we find

$$g^N = I_N .$$

We also obtain

$$h^N = I_N .$$

(ii) Since

$$1 + \omega + \omega^2 + \cdots + \omega^{N-1} = 0$$

we find

$$\text{tr } g = 0 .$$

(iii) Obviously we have

$$J^*_{(m_1,m_2)} = J_{(-m_1,-m_2)} \, .$$

(iv) Using equation (1) we find

$$J_{\mathbf{m}} J_{\mathbf{n}} = \omega^{\mathbf{n} \times \mathbf{m}/2} J_{\mathbf{m}+\mathbf{n}} \, .$$

(v) Using the result from (iv) we obtain

$$[J_{\mathbf{m}}, J_{\mathbf{n}}] = -2i \sin\left(\frac{2\pi}{N} \mathbf{m} \times \mathbf{n}\right) J_{\mathbf{m}+\mathbf{n}} \, .$$

Chapter 13

Graphs and Matrices

A *graph* consists of a non-empty set of points, called *vertices* (singular: vertex) or nodes, and a set of lines or curves, called *edges*, such that every edge is attached at each end to a vertex. We assume that $V = \{1, 2, \ldots, n\}$. A *digraph* (directed graph) is a diagram consisting of points, called vertices, joined by directed lines, called *arcs*. Thus each arc joins two vertices in a specified direction. To distinguish them from undirected graphs the edges of a digraph are called arcs.

The most frequently used representation schemes for graphs and digraphs are adjacency matrices. The *adjacency matrix* of an n-vertex graph $G = (V, E)$ is an $n \times n$ matrix A. The adjacency matrix $A(G)$ of a graph G with n vertices is an $n \times n$ matrix with the matrix element a_{ij} being the number of edges joining the vertices i and j.

The adjacency matrix $A(D)$ of a directed graph D with n vertices is an $n \times n$ matrix with the matrix element a_{ij} being the number of arcs from vertex i to vertex j.

A *walk* of length k in a graph is a succession of k edges joining two vertices. Edges can occur more than once in a walk. A *trail* is a walk in which all the edges (but not necessary all the vertices) are distinct. A *path* is a walk in which all the edges and all the vertices are distinct.

An *Eulerian graph* is a connected graph which contains a closed trail which includes every edge. The trail is called an Eulerian trail.

Problem 1. A walk of length k in a digraph is a succession of k arcs joining two vertices. A trail is a walk in which all the arcs (but not necessarily all the vertices) are distinct. A path is a walk in which all the arcs and all the vertices are distinct. Show that the number of walks of length k from vertex i to vertex j in a digraph D with n vertices is given by the ijth element of the matrix A^k, where A is the adjacency matrix of the digraph.

Solution 1. We use *mathematical induction*. Assume that the result is true for $k \leq K - 1$. Consider any walk from vertex i to vertex j of length K. Such a walk consists of a walk of length $K - 1$ from vertex i to a vertex p which is adjacent to vertex j followed by a walk of length 1 from vertex p to vertex j. The number of such walks is $(A^{K-1})_{ip} \times A_{pj}$. The total number of walks of length k from vertex i to vertex j will then be the sum of the walks through any p, i.e.

$$\sum_{p=1}^{n} (A^{K-1})_{ip} A_{pj} .$$

This is just the expression for the ijth element of the matrix $A^{K-1}A = A^K$. Thus the result is true for $k = K$. The result is certainly true for the walks of length 1, i.e. $k = 1$ since this is the definition of the adjacency matrix A. Therefore it is true for all k.

Problem 2. Consider a digraph. The out-degree of a vertex v is the number of arcs incident from v and the in-degree of a vertex V is the number of arcs incident to v. Loops count as one of each.

Determine the in-degree and the out-degree of each vertex in the digraph given by the adjacency matrix

$$A = \begin{pmatrix} 0 & 1 & 0 & 0 & 0 \\ 0 & 0 & 1 & 0 & 0 \\ 1 & 0 & 0 & 0 & 1 \\ 0 & 0 & 1 & 0 & 0 \\ 0 & 0 & 0 & 1 & 0 \end{pmatrix}$$

and hence determine if it is an Eulerian graph. Display the digraph and determine an Eulerian trail.

Solution 2. The out-degree of each vertex is given by adding the entries in the corresponding row and the in-degree by adding the entries in the corresponding column.
Vertex 1 has out-degree 1 and in-degree 1
Vertex 2 has out-degree 1 and in-degree 1

Vertex 3 has out-degree 2 and in-degree 2
Vertex 4 has out-degree 1 and in-degree 1
Vertex 5 has out-degree 1 and in-degree 1
Hence each vertex has out-degree equal to in-degree and the digraph is Eulerian. An Eulerian trial is, for instance, 1235431.

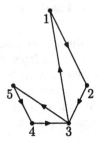

Problem 3. A digraph is strongly connected if there is a path between every pair of vertices. Show that if A is the adjacency matrix of a digraph D with n vertices and B is the matrix

$$B = A + A^2 + A^3 + \cdots + A^{n-1}$$

then D is strongly connected iff each non-diagonal element of B is greater than 0.

Solution 3. We must show both "if each non-diagonal element of B is greater than 0 then D is strongly connected" and "if D is strongly connected then each non-diagonal element of B is greater than 0".

Firstly, let D be digraph and suppose that each non-diagonal element of the matrix we have $b_{ij} > 0$, i.e. $b_{ij} > 0$ for $i \neq j$ and $i, j = 1, 2, \ldots, n$. Then $(A^k)_{ij} > 0$ for some $k \in [1, n-1]$, i.e. there is a walk of some length k between 1 and $n-1$ from every vertex i to every vertex j. Thus the digraph is strongly connected.
Secondly, suppose the digraph is strongly connected. Then, by definition, there is a path from every vertex i to every vertex j. Since the digraph has n vertices the path is of length no more than $n-1$. Hence, for all $i \neq j$, $(A^k)_{ij} > 0$ for some $k \leq n-1$. Hence, for all $i \neq j$, we have $b_{ij} > 0$.

Problem 4. Write down the adjacency matrix A for the digraph shown. Calculate the matrices A^2, A^3 and A^4. Consequently find the number of walks of length 1, 2, 3 and 4 from w to u. Is there a walk of length 1, 2, 3, or 4 from u to w? Find the matrix $B = A + A^2 + A^3 + A^4$ for the digraph and hence conclude whether it is strongly connected. This means finding out whether all off diagonal elements are nonzero.

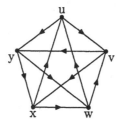

Solution 4. We have

$$A = \begin{array}{c} \begin{array}{ccccc} u & v & w & x & y \end{array} \\ \begin{pmatrix} 0 & 1 & 0 & 0 & 1 \\ 0 & 0 & 0 & 1 & 1 \\ 1 & 1 & 0 & 0 & 0 \\ 1 & 0 & 1 & 0 & 0 \\ 0 & 0 & 1 & 1 & 0 \end{pmatrix} \end{array}.$$

Thus the powers of A are

$$A^2 = \begin{pmatrix} 0 & 0 & 1 & 2 & 1 \\ 1 & 0 & 2 & 1 & 0 \\ 0 & 1 & 0 & 1 & 2 \\ 1 & 2 & 0 & 0 & 1 \\ 2 & 1 & 1 & 0 & 0 \end{pmatrix}, \quad A^3 = \begin{pmatrix} 3 & 1 & 3 & 1 & 0 \\ 3 & 3 & 1 & 0 & 1 \\ 1 & 0 & 3 & 3 & 1 \\ 0 & 1 & 1 & 3 & 3 \\ 1 & 3 & 0 & 1 & 3 \end{pmatrix}$$

$$A^4 = \begin{pmatrix} 4 & 6 & 1 & 1 & 4 \\ 1 & 4 & 1 & 4 & 6 \\ 6 & 4 & 4 & 1 & 1 \\ 4 & 1 & 6 & 4 & 1 \\ 1 & 1 & 4 & 6 & 4 \end{pmatrix}.$$

The number of walks from w to u is given by the $(3,1)$ element of each matrix so there is 1 walk of length 1, 0 walks of length 2, 1 walk of length 3 and 6 walks of length 4 from w to u. Walks from u to w are given by the $(1,3)$ element of each matrix. Thus there are 0 walks of length 1, 1 walk of length 2, 3 walks of length 3 and 1 walk of length 4 from u to w. The matrix B is given by

$$B = A + A^2 + A^3 + A^4 = \begin{pmatrix} 7 & 8 & 5 & 4 & 6 \\ 5 & 7 & 4 & 6 & 8 \\ 8 & 6 & 7 & 5 & 4 \\ 6 & 4 & 8 & 7 & 5 \\ 4 & 5 & 6 & 8 & 7 \end{pmatrix}.$$

Therefore the digraph is strongly connected (all the off diagonal elements are nonzero).

Bibliography

Aldous J. M. and Wilson R. J.
Graphs and Applications: An Introductory Approach
Springer Verlag (2000)

Bronson, R.
Matrix Operations
Schaum's Outlines, McGraw-Hill (1989)

Fuhrmann, P. A.
A Polynomial Approach to Linear Algebra
Springer-Verlag, New York (1996)

Golub, G. H. and Van Loan C. F.
Matrix Computations, Third Edition,
Johns Hopkins University Press (1996)

Grossman S. I.
Elementary Linear Algebra, Third Edition
Wadsworth Publishing, Belmont (1987)

Horn R. A. and Johnson C. R.
Topics in Matrix Analysis
Cambridge University Press (1999)

Johnson D. L.
Presentation of Groups
Cambridge University Press (1976)

Kedlaya K. S., Poonen B. and Vakil R.
The William Lowell Putnam Mathematical Competition 1985–2000,
The Mathematical Association of America (2002)

Lang S.
Linear Algebra
Addison-Wesley, Reading (1968)

Miller W.
Symmetry Groups and Their Applications
Academic Press, New York (1972)

Schneider H. and Barker G. P.
Matrices and Linear Algebra,
Dover Publications, New York (1989)

Steeb W.-H.
Matrix Calculus and Kronecker Product with Applications and C++ Programs
World Scientific Publishing, Singapore (1997)

Steeb W.-H.
Continuous Symmetries, Lie Algebras, Differential Equations and Computer Algebra
World Scientific Publishing, Singapore (1996)

Steeb W.-H.
Hilbert Spaces, Wavelets, Generalized Functions and Quantum Mechanics
Kluwer Academic Publishers, Dordrecht (1998)

Steeb W.-H.
Problems and Solutions in Theoretical and Mathematical Physics,
Second Edition, Volume I: Introductory Level
World Scientific Publishing, Singapore (2003)

Steeb W.-H.
Problems and Solutions in Theoretical and Mathematical Physics,
Second Edition, Volume II: Advanced Level
World Scientific Publishing, Singapore (2003)

Steeb W.-H., Hardy Y., Hardy A. and Stoop R.
Problems and Solutions in Scientific Computing with C++ and Java Simulations
World Scientific Publishing, Singapore (2004)

Van Loan, C. F.
Introduction to Scientific Computing: A Matrix-Vector Approach Using MATLAB,
Second Edition
Prentice Hall (1999)

Wybourne B. G.
Classical Groups for Physicists
John Wiley, New York (1974)

Index